Lecture Notes in Civil Engineering

Volume 2

Lecture Notes in Civil Engineering (LNCE) publishes the latest developments in Civil Engineering - quickly, informally and in top quality. Though original research reported in proceedings and post-proceedings represents the core of LNCE, edited volumes of exceptionally high quality and interest may also be considered for publication. Volumes published in LNCE embrace all aspects and subfields of, as well as new challenges in, Civil Engineering. Topics in the series include:

- Construction and Structural Mechanics
- Building Materials
- Concrete, Steel and Timber Structures
- Geotechnical Engineering
- Earthquake Engineering
- Coastal Engineering
- Hydraulics, Hydrology and Water Resources Engineering
- Environmental Engineering and Sustainability
- Structural Health and Monitoring
- Surveying and Geographical Information Systems
- Heating, Ventilation and Air Conditioning (HVAC)
- Transportation and Traffic
- Risk Analysis
- Safety and Security

More information about this series at http://www.springer.com/series/15087

Anastasios G. Sextos · George D. Manolis
Editors

Dynamic Response of Infrastructure to Environmentally Induced Loads

Analysis, Measurements, Testing, and Design

Springer

Editors
Anastasios G. Sextos
Department of Civil Engineering
University of Bristol
Bristol
UK

George D. Manolis
Department of Civil Engineering
Aristotle University of Thessaloniki
Thessaloniki
Greece

and

Department of Civil Engineering
Aristotle University of Thessaloniki
Thessaloniki
Greece

ISSN 2366-2557   ISSN 2366-2565   (electronic)
Lecture Notes in Civil Engineering
ISBN 978-3-319-56134-9   ISBN 978-3-319-56136-3   (eBook)
DOI 10.1007/978-3-319-56136-3

Library of Congress Control Number: 2017937271

Printed on acid-free paper

This Springer imprint is published by Springer Nature
The registered company is Springer International Publishing AG
The registered company address is: Gewerbestrasse 11, 6330 Cham, Switzerland

Preface

This book is the result of a three-year effort that paralleled the evolution of the research work conducted for the project entitled *DEutschland & GRIEchenland LABboratory (DeGrieLab): Hybrid and Virtual Experimentation for Infrastructure Lifecycle Maintenance and Natural Hazard Resilience* (www.degrielab.eu). This project was funded by the Deutscher Akademischer Austausch Dienst (DAAD) agency under contract number 57055451, and the actual duration was from January 1, 2014 to December 31, 2016. The project principal investigators from the German side were Professor Carsten Konke and Dr. Volkmar Zabel, Department of Civil Engineering, Bauhaus Universitat Weimar (BUW), in close collaboration with Professor Frank Wuttke who was the initial acting PI. Their counterparts from the Greek side were Associate Professor Anastasios G. Sextos (now serving a dual appointment with Department of Civil Engineering, Bristol University, Bristol, BS8 1TR, UK) and Prof. George D. Manolis, Department of Civil Engineering, Aristotle University (AUTH).

A number of international scientific events were organized in the framework of the above project, namely a workshop on Dynamic Analysis, Testing and Design of Infrastructure to Environmental Loads in Thessaloniki (November, 11–13/2014), followed by a same topic workshop held at Bauhaus Universitat Weimar (August, 24–26/2015) and a final workshop organized in Chalkidiki, Greece, on Recent Developments in Structural Health Monitoring for a Resilient Infrastructure (June, 31/5–3/2016). Two summer schools were also organized in Weimar and Thessaloniki in 2014 and 2015, respectively. The educational and research presentation material as well as 60 videos of the talks has all been made publically available in the project educational channel in YouTube (shortcut: https://tinyurl.com/jjzxlxa).

This book includes invited chapters by the participants of the last two workshops and encompasses five broad areas in civil engineering infrastructure, namely (a) structural health monitoring of bridges, (b) seismic excitation, monitoring, and response of buildings, (c) soil-structure interaction, (d) numerical methods, and (e) hybrid and experimental mechanics. The chapters are authored by project partners of the *DeGrieLab* project and colleagues from Europe and North America and can be classified as follows.

Chapters 1 and 2 deal with issues regarding the seismic response assessment of R/C bridges. Chapters 3–5 discuss bridge testing in terms of hybrid simulations, open field tests, and bridge component testing on an intercontinental scale through Internet linking. Chapter 6 is on the experimental assessment of isolators for existing R/C bridges, while Chap. 7 examines the mechanical response and numerical analysis issues for base isolators used in buildings. Chapters 8–10 present material on the various experimental testing possibilities for earthquake engineering purposes and component testing on shaking tables, on structural imaging techniques, and on structural health monitoring using wireless sensors, respectively. Chapters 11–14 focus on soil-structure interaction issues, energy-based methods, and inelastic dynamic analysis methods for R/C bridges, buried tunnels, buildings, and bridge piers, respectively. Finally, Chap. 15 closes with the quantification of seismic collapses of buildings.

This book is intended for practicing engineers, researchers, and graduate students. We hope that such readers will find this book useful for their work.

Finally, the editors would like to thank Mrs. Kleoniki Kyrkopoulou, MA, for her precious contribution in everyday management of the project as well as for proofreading the contents of the book. Thanks are also due to all who contributed and worked under this project, namely faculty, students, and administrators, as well as the Deutscher Akademischer Austausch Dienst (DAAD) agency for its generous support. We believe that this project contributed toward strengthening the academic ties between Greece and Germany and promoted mutual understanding and long-term collaboration among the partners involved.

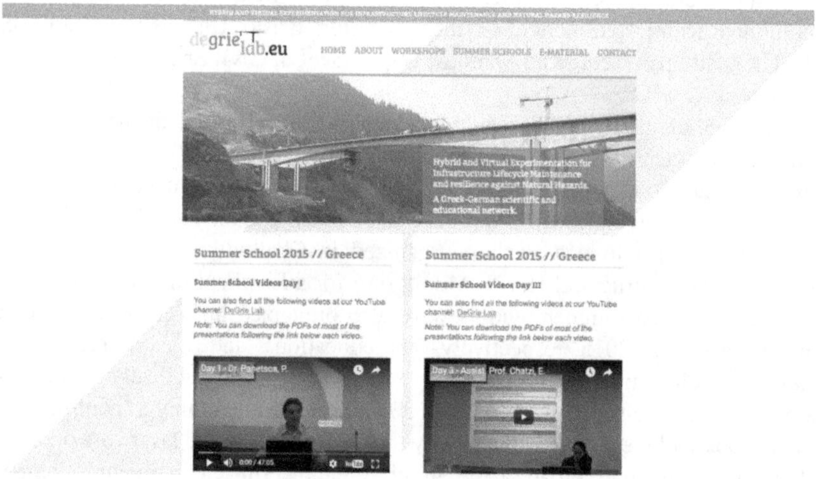

DeGrie Lab Project Educational Portal

DeGrie Lab Workshop, Thessaloniki, Greece (November, 11–13/2014)

DeGrie Lab Workshop, Weimar, Germany (August, 24–26/2015)

DeGrie Lab Workshop, Weimar, Germany (August, 24–26/2015)

Thessaloniki, Greece

Anastasios G. Sextos
George D. Manolis

Contents

1   Influence of Seismic Wave Angle of Incidence Over
    the Response of Long Curved Bridges Considering
    Soil-Structure Interaction   1
    Anastasios G. Sextos and Olympia Taskari

2   Alternative Approaches to the Seismic Analysis
    of R/C Bridges   19
    Tatjana Isakovic and Matej Fischinger

3   Multi-platform Hybrid (Experiment-Analysis) Simulations   37
    Oh-Sung Kwon

4   Field Structural Dynamic Tests at the Volvi-Greece
    European Test Site   65
    George C. Manos

5   An Intercontinental Hybrid Simulation Experiment
    for the Purposes of Seismic Assessment of a Three-Span
    R/C Bridge   77
    Anastasios G. Sextos and Olympia Taskari

6   Experimental Seismic Assessment of the Effectiveness
    of Isolation Techniques for the Seismic Protection
    of Existing RC Bridges   89
    L. Di Sarno and F. Paolacci

7   Modeling of High Damping Rubber Bearings   115
    Athanasios A. Markou, Nicholas D. Oliveto
    and Anastasia Athanasiou

8   Experimental Methods and Activities in Support
    of Earthquake Engineering   139
    Stathis N. Bousias

9   Time Reversal and Imaging for Structures   159
     C.G. Panagiotopoulos, Y. Petromichelakis and C. Tsogka

10   Decentralized Infrastructure Health Monitoring Using
     Embedded Computing in Wireless Sensor Networks   183
     Kosmas Dragos and Kay Smarsly

11   Effects of Local Site Conditions on Inelastic Dynamic
     Response of R/C Bridges   203
     Ioanna-Kleoniki Fontara, Magdalini Titirla, Frank Wuttke,
     Asimina Athanatopoulou, George D. Manolis
     and Petia S. Dineva

12   BEM-FEM Coupling in the Time-Domain for Soil-Tunnel
     Interaction   219
     S. Parvanova, G. Vasilev, P.S. Dineva and Frank Wuttke

13   Energy Methods for Assessing Dynamic SSI Response in
     Buildings   237
     Mourad Nasser, George D. Manolis, Anastasios G. Sextos,
     Frank Wuttke and Carsten Könke

14   Numerical and Experimental Identification
     of Soil-Foundation-Bridge System Dynamic Characteristics   259
     P. Faraonis, Frank Wuttke and Volkmar Zabel

15   Quantification of the Seismic Collapse Capacity
     of Regular Frame Structures   269
     David Kampenhuber and Christoph Adam

Chapter 1
Influence of Seismic Wave Angle of Incidence Over the Response of Long Curved Bridges Considering Soil-Structure Interaction

Anastasios G. Sextos and Olympia Taskari

Abstract Scientific research has shown that soil-structure interaction (SSI) should be investigated especially in the case of bridges with great importance or specific soil and structural characteristics. The most efficient way available nowadays to account for this phenomenon is by modeling the performance of soil, structure and foundation as a whole in the time domain. On the other hand, especially for the case of long, curved bridges, the issue of deciding a 'reasonable' incoming wavefield angle of incidence has not yet been scrutinized. Along these lines, the scope of this paper is to investigate the influence of the excitation direction of seismic motion in the case of long, curved bridges, using the most refined finite element model practically affordable in terms of computational cost. For this purpose, the long (640 m) and curved (R = 488 m) Krystallopigi Bridge was modeled using the finite element program ANSYS accounting for SSI both at the location of piers and abutments. The parametric study of different ground motion scenarios performed, highlights the complexity of the phenomenon and the difficulty in determining a 'critical' angle of excitation for all response quantities and all piers at the same time especially when soil-structure interaction is considered. Moreover, the dispersion of the results obtained indicates that the impact of ignoring that phenomenon and the role played by SSI effects may be significant under certain circumstances.

Keywords Bridges · Soil-structure interaction · Angle of excitation

A.G. Sextos (✉) · O. Taskari
Division of Structural Engineering, Department of Civil Engineering,
Aristotle University, 54124 Thessaloniki, Greece
e-mail: asextos@civil.auth.gr

A.G. Sextos
Department of Civil Engineering, University of Bristol Queens Building,
University Walk, Bristol BS8 1TR, UK

© Springer International Publishing AG 2017
A.G. Sextos and G.D. Manolis (eds.), *Dynamic Response of Infrastructure to Environmentally Induced Loads*, Lecture Notes in Civil Engineering 2,
DOI 10.1007/978-3-319-56136-3_1

1.1 Introduction

It is widely accepted (Pender 1993; Wolf 1994; Mylonakis et al. 2002) that soil–structure interaction (SSI) may significantly modify the dynamic characteristics of a structural system leading to a completely different (elastic or inelastic) dynamic behavior compared with the one expected when considering fixed support conditions. The reason is that there is significant variation between the foundation and the free-field earthquake motion, while since the foundation of a structure is never completely rigid, the soil flexibility affects the overall response of the structural system, depending on the relative rigidity and mass of the soil-foundation-superstructure system studied. Moreover, the fundamental period of vibration of the flexible-supported structure is elongated compared to the fixed-base case. This leads to lower spectral accelerations, and subsequently lower internal forces for all (i.e. lateral load and response spectrum) analyses that are performed on the basis of the code elastic or design spectrum and for structures with fundamental period corresponding to the plateau or the descending branch of the above spectra. The dynamic response of the flexibly supported system however under actual earthquake loading (i.e. with an arbitrary response spectrum) does not necessarily lead to reduced seismic demand (Mylonakis and Gazetas 2000; Sextos et al. 2003). In fact, the beneficial or detrimental effect of SSI on the dynamic response of a structure depends on a series of parameters such as the intensity of ground motion, the dominant wavelengths, the angle of incidence of the seismic waves, the chromatography, the stiffness and damping of soil as well as the size, geometry, stiffness, slenderness and dynamic characteristics of the structure itself. Finally, both the damping radiated at the soil-foundation interface and the hysteretic soil damping increase the amount of energy absorbed in the case of the flexibly supported system, compared to the fixed-base system which is only associated to the damping of the superstructure. Given the above features of SSI, the problem is often uncoupled into two, simultaneous but distinct mechanisms: the first is related to the interaction of the foundation with the surrounding deformable soil and the subsequent filtering of seismic motion (kinematic interaction) while the latter is associated to the dynamic stiffness of the superstructure given the incoming (filtered) foundation input motion and the additional inertial forces that are imposed to the foundation due to the vibration of the superstructure and the radiated wavefield (inertial interaction).

Due to the complexity of the problem, the most common approach to take SSI into consideration is through the uncoupling of the two main components of interaction, i.e. kinematic and inertial, and their subsequent superpositioning. Within this framework the foundation and the surrounding soil are replaced with appropriate for each stage springs and dashpots (Gazetas 1991; Makris and Gazetas 1992) while the response of the foundation itself is considered as the input motion for the (similarly supported) superstructure. On the other hand, modeling of the performance of soil, structure and foundation as a whole in the time domain using appropriate finite or boundary elements has always been a tempting approach. However, the inherent uncertainty in the spatial distribution of soil characteristics and earthquake ground motion as well as the computational cost related to modeling the propagation of seismic waves and ensuring appropriate stress distribution around the bridge foundation most often prohibit the development of large-scale finite element models for the study of the particular phenomenon. As a result, only few attempts to analyze the full soil-structure system have been performed so far for large bridge-soil systems (i.e. Humboldt Bay Bridge, Yan et al. 2004; Zhang et al. 2004; Meloland Road Over-crossing Bridge, Kwon and Elnashai 2006) while such a 'holistic' finite element approach is still deemed unrealistic for any practical purposes.

Additionally, research has shown that except for the SSI effect and the frequency content of the input motion, the direction of the excitation may significantly affect the dynamic response of irregular structures in the time domain. Despite the fact that numerous researchers have studied the importance of the excitation angle in the dynamic response of structures, the effect of soil compliance and damping on the relative sensitivity of a bridge to the direction of its excitation is typically not studied. As a result, the conclusions drawn based on theoretical approaches (such that of Penzien and Watabe 1975) or response spectrum analysis (Wilson and Button 1982; Smeby and Der Kiureghian 1985; Lopez and Torres 1997; Lopez et al. 2000; Hernandez and Lopez 2002), linear time history analysis (Athanato-poulou 2005) or on nonlinear analysis in the time domain (MacRae and Mattheis 2000; Tezcan and Alhan 2001; Rigato and Medina 2007) or pushover analysis (Moschonas and Kappos 2013) cannot be easily extrapolated for the case of large soil-foundation-bridge systems. Torsional sensitivity of a structure is an additional parameter that adds to the complexity of the problem as for some irregular buildings it has been shown (Tso and Smith 1999; Sextos et al. 2005; Aziminejad and Moghadam 2006) that the importance of the adopted direction of seismic excitation is strongly coupled with the contribution of the excited torsional modes of vibration and the subsequent nonlinear response of the structure. Moreover, the current seismic design framework (i.e. EC8-Part2 CEN 2005 and Greek code E39/99 Ministry of Public Works 1999 referring to bridges) is unclear as to the principal axes of excitation especially for the case of curved bridges. Hence, the designer cannot easily quantify the uncertainty related to the selection of the 'appropriate' direction of base excitation, which at the end and given the overall uncertainty, is usually assumed parallel and perpendicular to the chord for the two seismic input components respectively.

Therefore, the scope of this paper is to utilize the currently available computational capabilities in order to:

- investigate the effect of the excitation direction in the case of long, curved bridges, using the most refined finite element model practically affordable in terms of computational cost which includes the modeling of soil both at the location of piers and abutments,
- quantitatively evaluate the potential influence of the excitation direction of seismic motion in the case of long, curved bridges in terms of deck displacements and action effects,

- evaluate the scatter of the action effects as the direction of excitation is gradually modified,
- evaluate if the scatter depends on the soil-structure interaction, or the common assumption of a fixed base system does not affect the influence of the excitation angle in the case of curved bridges,
- investigate if the critical excitation angle coincides with the principal axes of a curved bridge (parallel and perpendicular to the bridge chord) and if not, to evaluate the consequences of the no clear instructions provided by the current codes regarding the direction of enforced excitation.

For this purpose, the long and curved Krystallopigi Bridge described below was adopted for study utilizing a large soil volume and investigating parametrically various scenarios of seismic wave angle of incidence. The description of the bridge configuration as well as the comparative results of the direction of excitation impact ignoring or considering soil-structure interaction are presented in the following.

1.2   Overview of the Bridge Studied

The Krystallopigi Bridge is a long curved structure that crosses a valley, as a part of the EGNATIA highway in West Macedonia region in Greece studied in detail by Paraskeva et al. (2006) while its response has been also evaluated for spatially variable earthquake ground motions (Sextos and Kappos 2008). The structure comprises of two curved but parallel sections; however, this study focuses on the left branch of the bridge which is a twelve span structure of a total length of 640 m. The two outer spans of this branch have a 44 m length each while the ten inner spans have a 55 m length. The curvature radius is equal to 488 m and its deck width is 13 m. The slope of the structure along the bridge axis varies (from 2.9 to 5.12%) while the deck transverse slope is constant and equal to 6%. A prestressed concrete box girder section is used for the deck while the piers consist of rectangular hollow reinforced concrete sections which in the pier top range are formed as solid rectangular sections for practical reasons (e.g. anchorage of prestressing cables). The structure is supported on eleven piers (M1–M11) of height that varies between 11 and 27 m. For the end piers (M1, M2, M3, M9, M10, M11) a bearing type pier-to-deck connection was adopted, allowing movement in the longitudinal direction but restricting movement in the transverse direction, while the interior piers were constructed as monolithically connected to the deck. Foundation soils are in general composed of soft ($v_s = 250$ m/s) to moderate stiffness ($v_s = 400$ m/s) layers, as well as stiff limestone formations ($v_s = 1800$ m/s). A number of piers are supported on groups of piles while others on surface foundations; their configuration and length depend on the foundation soil properties. Specifically, the abutment A1 as well as piers M1–M9 are supported on 1.2 m diameter group of piles which cross the surface clay layer up to the level of submerged limestone while the

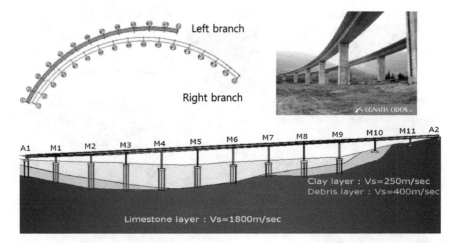

Fig. 1.1 Layout of the bridge configuration

piers M10–M11 as well as the abutment A2 is directly founded on the stiff lime-stone outcrop. The layout of bridge configuration is presented in Fig. 1.1.

1.3 Modeling of the Bridge-Soil System

Both the Krystallopigi Bridge and the near field soil were modeled in 3-Dimensions using the finite element program ANSYS (ANSYS Inc. 2006a) and the inherent language APDL (ANSYS Inc. 2006b). The particular programming approach elaborates the reversibility of the finite element model developed and the effective management of the post-processing data resulting from the parametric analyses performed.

The deck and the piers, which sections vary along the bridge and the piers axis respectively, were modeled with 3-Dimensional beam elements. A dense grid of beam elements was generated at the pier-deck connection range as well as at locations of abrupt deck or pier section dimensions. As a result, 220 beam elements were used for the deck discretization while 8–10 beam elements were used for each middle pier. It is noted that the line (beam) pier elements were connected to the (solid element) supporting pile cap through appropriate coupling equations in order to ensure realistic stress transmission and distribution as illustrated in Fig. 1.2 (Sextos et al. 2004).

As already mentioned, the foundation soil is composed of clay, debris and limestone layers of different height along the bridge. In order to be able to uncouple the relative impact of the excitation direction from the inherent uncertainty related to the variation of soil properties with soil depth and bridge length, the soil domain was simulated as homogenous and characterized by a uniform mean value of modulus of elasticity that was taken equal to 30 MPa. A 700 m × 240 m soil

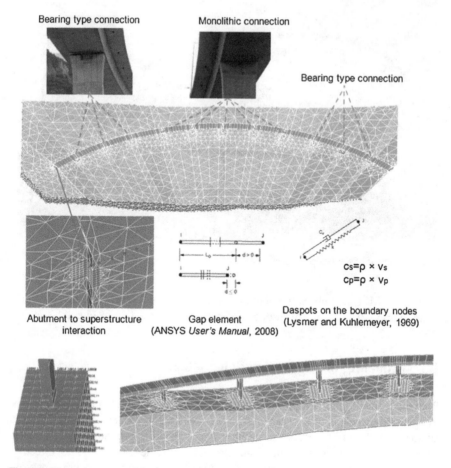

Fig. 1.2 Finite element model of the Krystallopigi bridge

domain was generated with depth varying from 2 to 40 m depending on the actual topography. The need to incorporate the exact pile group configuration and length below each pier and to model soil-to-pile and pile-to-pile dynamic interaction, essentially determined the finite element mesh geometry at the vicinity of the bridge. As a result, a dense grid of 8-node brick elements (Solid185 ANSYS type, 3 DOF per node) was adopted for an area twice as large the pile group dimensions, gradually leading to a coarse mesh grid of 20-node solid elements (Solid186 ANSYS type, 3 DOF per node).

During the finite element modeling of the system an effort was made to incorporate the effect of embankment-abutment-superstructure interaction since it has been shown both numerically and through measurements from bridges in California that not only the stiffness and damping of the system is strongly affected but also the incoming input motion maybe significantly amplified (Kotsoglou and Panta-zopoulou 2006). Along these lines, the embankment of the left abutment (i.e. A1) was modeled in detail along a critical length of 50 m. It is noted that there was no reason to replicate the aforementioned discretization approach for the right abutment (i.e. A2) as well since, according to the geotechnical study available, it was founded on stiff limestone formations. The superstructure to abutment interaction on the other hand, was taken into account with the use of appropriate gap elements (ANSYS type Link10) while full contact was assumed between abutment and embankments. As noted previously, 3-Dimensional beam elements were used for modeling the foundation piles, the length of which essentially coincides with the dimensions of soil mesh at the vicinity of the pile group.

As for the boundary conditions, appropriate dashpots with values depending on soil characteristics were implemented on the lateral surfaces of the soil domain in order to diminish reflections of waves on the particular boundaries (Lysmer and Kuhlemeyer 1969) with the exception of the area of abutment A2 due to the aforementioned physical restraint provided by the supporting stiff outcrop. Similarly, the base of the model was also fixed to elaborate uniform acceleration earthquake input of the system for various angles of enforced base excitation.

As it is well known, higher frequencies and mode shapes of the spatially discretized equations generally do not accurately represent the dynamic response of such a complex system in the framework of transient analysis. Moreover, filtering of high frequencies is not always accurately performed while algorithmic damping provided by Newmark's method often leads to a lower level of accuracy (Belytschko and Hughes 1983). For this reason, the Hilber-Hughes-Taylor integration method was used with an integration constant $\alpha = -0.30$ in order to obtain an unconditionally stable, second order accurate scheme. Rayleigh damping was also implemented with appropriate constants (a = 0.91 and b = 0.003) so that the overall system damping in the range of frequencies of interest would vary between 5 and 10%.

1.4 Effect of Direction of Excitation on the Seismic Response of Bridges

1.4.1 Seismic Scenarios Studied

As a means to quantify the effect of seismic motion incidence angle on the dynamic response of the bridge studied, two horizontal and perpendicular ground acceleration components were imposed simultaneously along the structure's axes by assuming a gradual rotation of the excitation vector around the z-z (vertical) axis. For this purpose, four different earthquake events were investigated based on the records obtained from the Kozani, Athens, Lefkada and Thessaloniki earthquake (Table 1.1). The response spectra of the input motions are presented in Fig. 1.3. The base excitation (i.e. at the base of the soil medium) was computed after appropriate deconvolution and baseline correction process for both the horizontal

Table 1.1 Seismic excitation scenarios (before deconvolution process)

SCEN	Seismic event	Magnitude	Record position	PHA (m/s²)
1	Kozani (1995)	5.2 (Mw)	Prefecture	2.05
2	Athens (1999)	6.0 (Mw)	Chalandri	1.58
3	Lefkada (2003)	6.4 (Mw)	Hospital	4.12
4	Thessaloniki (1978)	5.12 (Ms)	City hotel	1.43

Fig. 1.3 Response spectra of the earthquake records used for the analyses

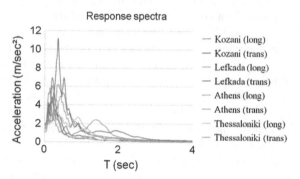

components. Site response, in terms of the amplitude and frequency amplification of seismic waves as they propagate through the soil medium, were taken into account inherently by the structure of the 3-Dimensional soil domain, hence, no specific analysis had to be performed for this purpose. It is only noted that for simplicity, soil was considered as linear elastic; consequently, as it is shown in the following, it was the relative effect of angle of incidence that was studied and no absolute values of displacements or bending moments were directly compared.

1.4.2 Methodology

Having modeled the soil-foundation-structure system and generated the earthquake input motion scenarios a parametric scheme was adopted, employing different angles of base excitation (i.e. 0° to 180° at a step of 15°) for the four aforementioned seismic events.

For the sake of comparison, a reference excitation angle ($\theta = 0°$) was adopted corresponding to the simultaneous excitation along the chord and the perpendicular to the chord axes of the curved bridge under study. By obtaining the results of the reference analysis which henceforth will be named $\theta = 0$ analysis, the analyses are repeated for all scenarios for the alternative cases denoted as $\theta = i$ analysis, where i is the incidence angle of earthquake excitation with respect to the reference coordinate system (Fig. 1.4). Based on this assumption, the effect of the direction of excitation is expressed in terms of the orientation effect ratio $r(\theta_i)$ (Athanatopoulou 2005) for both the displacements and the member forces. Apparently, the particular

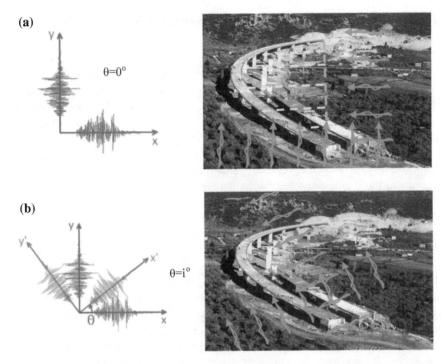

Fig. 1.4 Excitation under **a** the reference incidence angle of seismic wavefield ($\theta = 0°$), **b** a random incidence angle of seismic wavefield ($\theta = i°$)

ratio is equal to the deck absolute displacement or the pier base bending moment resulting from a $\theta = i$ analysis, divided to the deck absolute displacement or the pier base bending moment respectively for the $\theta = 0$ analysis:

$$r(\theta_i) = \frac{\max|R_{\theta \neq 0}(\theta_i, t)|}{\max|R_{\theta = 0}(t)|} \tag{1.1}$$

where i = 0° to 180° at a step of 15°, max|R$\theta \neq$ 0(θ, t)| is the maximum response value under $\theta \neq 0$ excitation and max|Rθ = 0(t)| is the maximum response value under $\theta = 0°$ excitation. A value of this ratio that exceeds 1.0 represents the unfavorable case of displacement or bending moment increase, while values that are less than 1.0 are deemed beneficial. After analyzing the five different seismic scenarios for the bridge-soil system, the orientation effect ratios of displacements and member forces were derived. From all the results obtained, the deck displacements which are parallel and perpendicular to the curved bridge chord at each pier location as well as the pier base bending moments around the strong pier section axis are presented and discussed herein.

1.4.3   Analyses Results of the Soil-Structure System

The deck displacement orientation effect ratios which are parallel and perpendicular
to the bridge chord at each pier and abutment location were computed for the
soil-structure system for the four input motions. It was observed that in many cases,
the maximum values of the ratios do not occur for the zero angle of incidence (i.e.
propagation parallel and perpendicular to bridge chord). It was also observed that
the excitation under an alternative angle of incidence (θ_i) may significantly affect in
an unfavorable way ($r(\theta_i) > 1$) a pier while the influence to an adjacent pier to be
favorable ($r(\theta_i) < 1$). Moreover the rotation of the wavefront plane at various angles
appears to have different effect to the parallel to the bridge chord deck displacements in comparison to the deck displacements which are perpendicular to the
chord. For instance, the values of the deck displacements orientation effect ratios
which are parallel to the bridge chord are in general less than unity, however the
ratios of the perpendicular to the bridge chord displacements are much greater.

The analyses results for the soil-structure system are presented in Figs. 1.5 and
1.6 for all the seismic excitation angles that were considered. Each radar type
diagram illustrates the value of the orientation effect ratio $r(\theta_i)$ for the (perpendicular to the bridge chord) deck displacements at the location of the piers and the
abutments for each excitation angle studied (i.e. 0° to 180° at a step of 15°). The
size of the radar type line essentially reflects the impact of different angles of
excitation (further from the center at values larger than 1.0 correspond to detrimental displacement increase) whereas the shape of the polygon shows whether the
observed increase is uniform for all angles or occurs for specific directions of base
excitation. For the sake of comparison the maximum value of 3.0 (maximum
showed value) was selected for all the diagram axes. At first it can be clearly noted

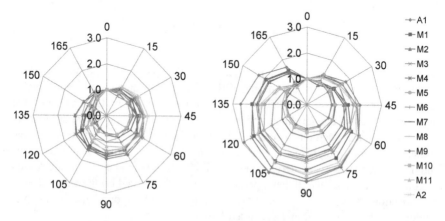

Fig. 1.5 Deck displacements orientation effect ratios at the location of piers and abutments which
are perpendicular to the bridge chord for the Kozani earthquake (*left*) and the Athens earthquake
(*right*)

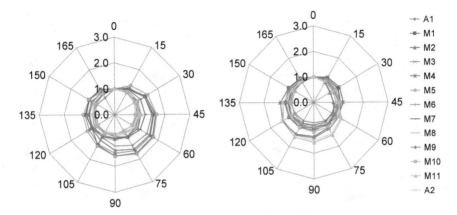

Fig. 1.6 Deck displacements orientation effect ratios at the location of piers and abutments which are perpendicular to the bridge chord for the Lefkada earthquake (*left*) and the Thessaloniki earthquake (*right*)

that the influence of the direction of excitation is different for each deck point examined but also different depending on the characteristics of the earthquake motion (scenario) studied. For instance (Figs. 1.5 and 1.6), the deck displacements at the position of pier M9 due to the Athens earthquake seems to be rather independent of the excitation angle and the reference earthquake used (effect ratio always close to 1.0), whereas displacement of the deck at the location of pier M4 present ratio which exceeds the value of 3.0 for particular excitation directions (i.e. 90°) and specific scenarios (Athens earthquake). As a consequence, a single value of a 'critical' angle of wave incidence or a 'critical' earthquake frequency content that leads uniformly to a global reduction or increase in displacements cannot be defined.

As an effort to quantify the influence of the excitation direction, the covariance of the orientation effect ratio was calculated for each pier for all the examined angles as well as the mean value of the covariance values of all the piers. The mean value of the covariance if seen as a gross measure or the error introduced when studying the particular bridge solely on the basis of two horizontal components along the chord and its perpendicular axis, that is ignoring the importance of the direction of excitation, is of the order of 0.20.

1.4.4 Analyses Results of the Fixed Base System

Although the SSI effect is proved to significantly affect the dynamic response of structures, this phenomenon is usually ignored by the practical engineering society. For this purpose, the aforementioned parametric analysis scheme was repeated for a fixed at the base of piers bridge as an effort to investigate whether this assumption

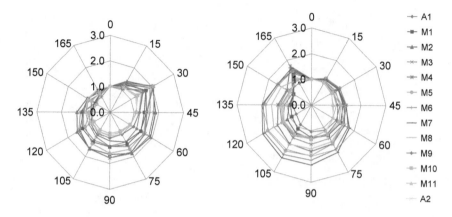

Fig. 1.7 Deck displacements orientation effect ratios at the location of piers and abutments which are perpendicular to the bridge chord for the Kozani earthquake (*left*) and the Athens earthquake (*right*)

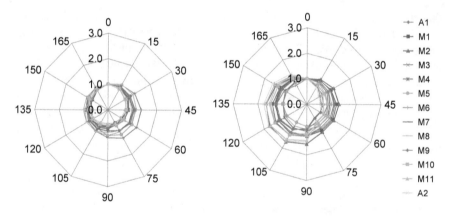

Fig. 1.8 Deck displacements orientation effect ratios at the location of piers and abutments which are perpendicular to the bridge chord for the Lefkada earthquake (*left*) and the Thessaloniki earthquake (*right*)

does modify the response patterns observed or the 'critical' (most detrimental) angle by rotating the vector of excitation as they obtained from the study including the entire soil volume beneath and around the bridge.

Figures 1.7 and 1.8 present the analyses results for the fixed-base system. It is interesting to notice that the same conclusion which was extracted for the soil-structure system can be drawn for the case of the fixed-base system. Again, the orientation effect ratio can be grater or lower to unity dependent on the angle of excitation studied, the action effect studied, the earthquake scenario and the location of the bridge. But in contrast to the refined finite element model that accounts for soil-foundation-bridge interaction, the critical combinations that lead to maximum

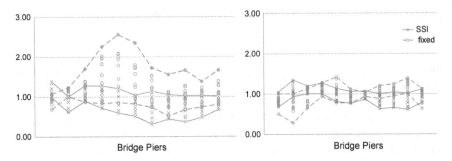

Fig. 1.9 Scatter of bending moments orientation effect ratios at piers of the SSI and the fixed base system for the Kozani earthquake (*left*) and the Athens earthquake (*right*)

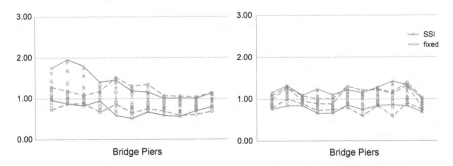

Fig. 1.10 Scatter of bending moments orientation effect ratios at piers of the SSI and the fixed base system for the Lefkada earthquake (*left*) and the Thessaloniki earthquake (*right*)

response values are completely different. As reason for such a distinct behavior between the flexibly and the rigidly fixed structure is that different dynamic characteristics of the curved bridge lead to different interplay patterns between the structure, the earthquake input and the direction of its application. The same conclusions were derived for the bending moments as well. The lack of a stable 'critical' angle of base excitation is clearly shown in Figs. 1.9 and 1.10 where the scatters as well as the envelopes of bending moments (around the strong pier section axis) are presented. It can be observed that the most detrimental angle of incidence varies in an unpredictable manner for all piers and all earthquake scenarios studied; an observation that is in agreement with previous studies (Athanatopoulou 2005).

Finally, it is interesting to notice that covariance of the order of 0.20 is also observed for the fixed system; however the individual set of orientation ratio $r(\theta_i)$ values that lead to the (same) covariance present a completely different distribution among the piers. As a result, it is deemed that not only the assumption of a single direction of excitation might hide significant aspects of the complex dynamic response of a curved bridge, but also even a refined approach in terms of direction of excitation that neglects the role played by the soil may be proven non conservative.

1.4.5 Interpretation of the Analyses Results

In an effort to interpret the changes in the seismic behavior of curved bridges subjected to the same input motion under different excitation angles, the transverse response of the M6 pier was selected for further study. The M6 pier is located at the middle of the bridge and the principal axes of its section coincide with the longitudinal and the transverse axis of the whole structure (x and y axis respectively). The Fourier spectrum of the response along the transverse bridge direction at the top of pier M6 was calculated for each examined angle of incidence. Figure 1.11 illustrates the Fourier spectra amplitude of deck displacements at the location of pier M6 for the 0° (reference), 15° and 30° excitation angles when the bridge was subjected to Lefkada earthquake. Apparently, the spikes of the spectra correspond to the natural frequencies of the bridge in which the pier M6 oscillates (also confirmed by the eigen analysis of the fixed base system). Particularly, the eigen analysis showed that the 1st (f_1 = 0.69 Hz) and the 3rd mode (f_3 = 1.34 Hz) are translational along the longitudinal axis of the bridge (x axis) while the 2nd (f_2 = 1.15 Hz) and the 4th modes (f_4 = 1.60 Hz) are translational along the transversal axis (y axis).

It is generally observed that the 1.15 Hz frequency dominates the response along the transverse direction of the Pier M6 while the other frequencies are activated in a smaller extent, thus they have a smaller influence to the overall pier oscillation. Specifically, if the θ = 15° excitation angle is examined, the ratio of the Fourier spectra amplitudes between the 15° and the reference excitation angle at the dominant frequency ($f(1.15\ \text{Hz})_{\theta=15}/f(1.15\ \text{Hz})_{\theta=0}$) is equal to 6.87/6.23 = 1.10; a value that coincides with the orientation effect ratio which had already been calculated (Fig. 1.8). Similarly, when the excitation under the 30° angle is examined ($f(1.15\ \text{Hz})_{\theta=30}/f(1.15\ \text{Hz})_{\theta=0}$) the ratio is equal to 7.42/6.23 = 1.20; a value that also coincides with the orientation effect ratio (Fig. 1.8). However there is a discrepancy between the Fourier spectra ratio and the orientation effect ratio as the angle of incident gradually increases from 0° to 90°. This is attributed to the fact that the rotation of the seismic wavefield around the vertical axis leads to the gradually excitation of the other modes which influence the pier M6 response.

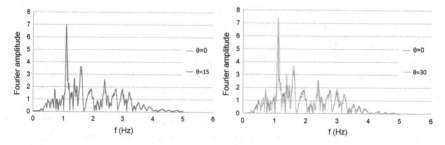

Fig. 1.11 Fourier spectra at the top of pier M6 for 0°, 15° and 30° angle of incidence (Lefkada earthquake)

Consequently, although the f = 1.15 Hz is the dominant mode, other modes are additionally activated, participate in the pier oscillation and therefore they affect the values of the orientation effect ratios.

Finally, it is noted that the same procedure cannot be easily applied for the case of the soil-structure system since such a refined model presents strong coupling between the soil and the bridge superstructure modes. Nevertheless, the interpretation of the bridge seismic response as the incidence angle gradually changes can be similar.

1.5   Conclusions

This paper aims to investigate the influence of the excitation direction of the seismic motion on the dynamic response of curved bridges (with emphasis in the long and curved in plan Krystallopigi Bridge), using a refined finite element model that accounts for the interaction between the approach embankment, the abutment, the surrounding soil, the foundation and the bridge structure. Through the parametric analysis performed that involved a set of five ground motion scenarios, the following conclusions were drawn:

- The typical analysis approach for the case of curved bridges in the time domain according to which a single direction of base excitation is adopted, may lead to non conservative estimates of seismic demand for specific piers and action effects and hide significant aspects the bridge complex response.
- Excitation of such a curved bridge along the $\theta = 0°$ or $90°$ is proven not necessarily the most critical, as one might have anticipated for the longitudinal and the transversal dynamic response respectively; a conclusion also in agreement with other researchers (Athanatopoulou 2005; Rigato and Medina 2007).
- The determination of the critical excitation angle that would lead to uniformly detrimental deck displacements or pier stress increase cannot be easily defined (or might even not exist at all) since the distribution of the orientation effect ratio along the piers for various angles of incidence and earthquake characteristics does not follow a predictable manner. However, one might be tempted to notice that this distribution is not completely random, since almost all resulting response increase polygons presented in Figs. 1.6, 1.7, 1.8, 1.9, 1.10 and 1.11 essentially resemble a circle with a center that is shifted from the origin. It is therefore feasible to claim that despite the overall problem uncertainty, the gradual modification of the excitation direction leads to equally gradual (and definitely not random) modification of the response.
- In order to obtain a reliable estimate of the maximum member forces and displacements of a curved bridge excited in the time domain, the designer has to perform analyses for various excitation angles between $0°$ and $180°$.

- It is unrealistic to adopt refined procedures regarding the ground motion characteristics and its angle of incidence if soil-structure-interaction effects are not modeled properly.
- It is considered that the effect of the excitation angle should be studied more thoroughly starting from the extrapolation of the parametric scheme described above, for the case of other equally realistic bridge configurations.

Acknowledgements The authors would like to thank EGNATIA ODOS for providing the necessary data for the superstructure-foundation-soil system of the Krystallopigi Bridge as well as their colleague Dr. Th. Paraskeva, whose previous modeling approach for the study of the particular bridge was used to validate the 3-Dimensional model presented herein.

References

ANSYS Inc. (2006a) ANSYS user's manual, Version 10.0. Canonsburg, PA

ANSYS Inc. (2006b) ANSYS programmers guide. Canonsburg, PA

Athanatopoulou A (2005) Critical orientation of three correlated seismic components. Eng Struct 27(2):301–312

Aziminejad A, Moghadam AS (2006) Behavior of asymmetric structures in near field ground motions. In: 1st European conference on earthquake engineering and seismology, Geneva, Switzerland, September

Belytschko T, Hughes T (1983) Computational methods for transient analysis. Elsevier Science Publishers, BV

CEN (Comité Européen de Normalisation) (2005) Eurocode 8: design provisions of structures for earthquake resistance—part 2: Bridges. prEN1998-2, Final draft, Brussels

Gazetas G (1991) Formulas & charts for impedance functions of surface and embedded foundation. J Geotech Eng-ASCE 177(9):1363–1381

Hernandez JJ, Lopez OA (2002) Response to three-component seismic motion of arbitrary direction. Earthq Eng Struct D 31(1):55–77

Kotsoglou A, Pantazopoulou S (2006) Modeling of embankment flexibility and soil-structure interaction in integral bridges. In: 1st European conference on earthquake engineering and seismology, Geneva, Switzerland, September

Kwon O-S, Elnashai AS (2006) Analytical seismic assessment of highway bridges with soil-structure interaction. In: 4th international conference on earthquake engineering, Taipei, Taiwan, October

Lopez OA, Torres R (1997) The critical angle of seismic incidence and the maximum structural response. Earthq Eng Struct D 26(9):881–894

Lopez OA, Chopra AK, Hernandez JJ (2000) Critical response of structures to multicomponent earthquake excitation. Earthq Eng Struct D 29(12):1759–1778

Lysmer J, Kuhlemeyer RL (1969) Finite dynamic model for infinite media. ASCE J Eng Mech Div Proc ASCE 95(EM4):859–877

MacRae GA, Mattheis J (2000) Three-dimensional steel building response to near-fault motions. J Struct Eng-ASCE 126(1):117–126

Makris N, Gazetas G (1992) Dynamic pile-soil-pile interaction. Part II: Lateral and seismic response. Earthq Eng Struct D 21(2):145–162

Ministry of Public Works (1999) Circular E39/99, Guidelines for the seismic analysis of bridges (in Greek)

Moschonas I, Kappos A (2013) Assessment of concrete bridges subjected to ground motion with an arbitrary angle of incidence: static and dynamic approach. B Earthq Eng 11(2):581–605

Mylonakis G, Gazetas G (2000) Seismic soil-structure interaction: beneficial or detrimental? J Earthq Eng 4(3):277–301

Mylonakis G, Gazetas G, Nikolaou S, Chauncey A (2002) Development of analysis and design procedures for spread footings. MCEER, Technical report 02-0003, US

Paraskeva T, Kappos A, Sextos A (2006) Extension of modal pushover analysis to seismic assessment of bridges. Earthq Eng Struct D 35(10):1269–1293

Pender MJ (1993) Aseismic pile foundation design analysis. Bull NZ Natl Soc Earthq Eng 26 (1):49–160

Penzien J, Watabe M (1975) Characteristics of 3-D earthquake ground motions. Earthq Eng Struct D 3:365–373

Rigato A-B, Medina R (2007) Influence of angle of incidence on seismic demands for inelastic single-storey structures subjected to bi-directional ground motions. Eng Struct 29(10):2593–2601

Sextos A, Pitilakis K, Kappos A (2003) Inelastic dynamic analysis of RC bridges accounting for spatial variability of ground motion, site effects and soil-structure interaction phenomena. Part 1: Methodology and analytical tools. Earthq Eng Struct D 32(4):607–627

Sextos A, Kappos A, Mergos P (2004) Effect of soil-structure interaction and spatial variability of ground motion on irregular bridges: the case of the Krystallopigi bridge. In: 13th world conference on earthquake engineering, Vancouver, Canada, August

Sextos A, Pitilakis K, Kirtas E, Fotaki V (2005) A refined computational framework for the assessment of the inelastic response of an irregular building that was damaged during the Lefkada earthquake. In: 4th European workshop on the seismic behaviour of irregular and complex structures, Thessaloniki, Greece, August

Sextos A, Kappos AJ (2008) Seismic response of bridges under asynchronous excitation and comparison with EC8 design rules. B Earthq Eng 7:519–545

Smeby W, Der Kiureghian A (1985) Modal combination rules for multi-component earthquake excitation. Earthq Eng Struct D 13:1–12

Tezcan SS, Alhan C (2001) Parametric analysis of irregular structures under seismic loading according to the new Turkish Earthquake Code. Eng Struct 23(6):600–609

Tso WK, Smith R (1999) Re-evaluation of seismic tensional provisions. Earthq Eng Struct D 28 (8):899–917

Wilson EL, Button M (1982) Three-dimensional dynamic analysis for multicomponent earthquake spectra. Earthq Eng Struct D 10(3):471–476

Wolf JP (1994) Foundation vibration analysis using simple physical models. Prentice Hall, New Jersey, USA

Yan L, Elgamal A, Yang Z, Conte JP (2004) Bridge-foundation-ground system: a 3D seismic response model. In: 3rd international conference on earthquake engineering, Nanjing, China, October

Zhang Y, Acero G, Conte J, Yan Z, Elgamal A (2004) Seismic reliability assessment of a bridge ground system. In: 13th world conference on earthquake engineering, Vancouver, Canada, August

Chapter 2
Alternative Approaches to the Seismic Analysis of R/C Bridges

Tatjana Isakovic and Matej Fischinger

Abstract Two alternative approaches for the design of R/C bridges are compared in this work, namely the traditional Strength Based Design (SBD) and the Direct Displacement Based Design (DDBD). It is found that these two methods give the same results when the same set of assumptions are employed. These are (a) the yield curvature (displacement) is nearly invariant for the chosen type of steel and geometry of the critical cross-section, (b) the equivalent pre-yielding stiffness is strongly correlated to the strength, and (c) the equal displacement rule is applied in both cases. The basic assumptions and properties behind the non-linear pushover-based methods, which are included in modern design codes, are reviewed and some specifics related to their use for the analysis of bridges are presented and briefly discussed.

Keywords Seismic analysis · Strength based design · Direct displacement based design · Pushover based analysis · Bridges · Yield displacements · Pre-yield stiffness · Effective stiffness

2.1 Introduction

Although that the response of most R/C bridges, which are subjected to strong earthquake load is predominantly nonlinear, their design is typically based on the results of elastic methods of analysis. For example, the Eurocode 8/2 standard (CEN 2005) defines the modal response spectrum analysis as the basic method of analysis. The acceleration design spectrum, used in this type of analysis, is typically reduced based on the chosen behaviour (reduction) factor for the bridge at hand.

T. Isakovic (✉) · M. Fischinger
Faculty of Civil and Geodetic Engineering, University of Ljubljana, Jamova 2, 1000 Ljubljana, Slovenia
e-mail: tatjana.isakovic@fgg.uni-lj.si

M. Fischinger
e-mail: matej.fischinger@fgg.uni-lj.si

© Springer International Publishing AG 2017
A.G. Sextos and G.D. Manolis (eds.), *Dynamic Response of Infrastructure to Environmentally Induced Loads*, Lecture Notes in Civil Engineering 2, DOI 10.1007/978-3-319-56136-3_2

This reduction defines the required strength of the structure. The larger reduction of forces, in general, means that the provided strength of the structure would be smaller. Taking into account the basic principle of the earthquake engineering, namely the equal displacement rule, further means that smaller provided strength should be accompanied by larger ductility capacity in the structure.

Recently, some doubts have been expressed about the validity of the equal displacement rule. A new design approach "Direct displacement based design"— DDBD was proposed (Priestley et al. 2007) as an alternative to the traditional "Strength Based Design"—SBD. Several opinions that DDBD is more economical than SBD have been presented (e.g., Martini 2007; Rahman and Sritharan 2011). In Sect. 2.2 a comparison of these methods is provided. Since both methods suppose that the response of structure is governed predominantly by one mode of vibration, they are compared considering only those structures, which can be modelled using single-degree-of-freedom (SDOF) model with reasonable accuracy. Comparison is limited to structures with fundamental periods of vibration in the constant velocity region of the spectrum.

In Sects. 2.2.1 and 2.2.2, a discussion about the basic assumptions of DDBD and the equal displacement rule is presented, respectively. The comparison of the methods is provided in Sect. 2.2.3 taking into account same basic assumptions, namely (a) the yield displacement is almost invariant for the chosen quality of the steel and the geometry of the structure, (b) the equivalent pre-yielding stiffness is strongly correlated to the strength, and (c) the equal displacement rule is applied in both cases. It means that both methods were modified as (a) in SBD the pre-yielding stiffness was estimated taking into account the basic assumption of DDBD method (the invariant yield displacement) and (b) in DDBD the equivalent damping was defined taking into account the basic assumption of SBD (the equal displacement rule). The modified methods are also compared with their original versions.

As previously mentioned, the seismic response of most RC bridges is non-linear. Thus the majority of the modern codes and design guidelines introduce the non-linear methods into design practice in order to estimate the seismic response more realistically. In general, the most refined and accurate inelastic method is the nonlinear response history analysis (NRHA). Nevertheless, it is only infrequently used in the design practice, since it is, for the moment, too complex for regular use by practising engineers. To simplify the nonlinear analysis and to make it more suitable for design practice, different static inelastic methods have been developed. Most of them are based on the pushover analysis. They are considered more user-friendly and relatively easy to understand. An overview of basic features of such methods, which are included in different codes, is provided in Sect. 2.3.

Most of the above methods have been primarily developed for the assessment of the seismic response of buildings. Since the response of bridges is often quite different from that of buildings, specific items that should be taken into account when these methods are applied to bridges are also briefly reviewed and discussed in Sect. 2.3.

2.2   Basic Concepts and Comparison of the DDBD and SBD

2.2.1   Basics of the DDBD

The DDBD is not a standard design tool. Therefore the main steps of the method are first reviewed (see "*The basic steps of DDBD*"). One of the basic assumptions of this method is that the yield curvature of critical cross-sections is almost invariant and depends only on the chosen quality of the reinforcing steel and chosen geometry of structural elements. This further implies that the yielding displacement does not depend on the strength of the structure, but only on its geometry. This assumption is discussed in sub-section "*Constant yield displacement*". The equal displacement rule is not included into DDBD. Since it is demonstrated in Sect. 2.2.3 that DDBD and SBD give the same results, when the same input data and the assumptions are used, the DDBD is modified at the end of this section (see *Modified DDBD taking into account equal displacement rule*) taking into account this basic principle of the seismic engineering.

The basic steps of the DDBD

The first step of DDBD is the estimation of the yield displacement Δ_y. It can be estimated in the way, described in the next sub-section. In the second step the ultimate displacement Δ_u is defined based on the yield displacement and the chosen ductility μ, or by taking into account drift limitations. If drift limitation is relevant, the ductility μ is calculated based on the ratio of the ultimate displacement (corresponding to drift limitation) and the yield displacement.

In the next step, the equivalent viscous damping at the peak response ξ_{eq} (corresponding to the ultimate displacement) is estimated, based on the type of the analysed structure and the type of used material. In the case of concrete bridges with effective periods longer than 1 s, the following equation has been proposed (Priestley et al. 2007):

$$\xi_{eq} = 0.05 + 0.444((\mu - 1)/\mu\pi) \qquad (2.1)$$

Then the equivalent viscous damping is used to reduce the displacement spectrum, which corresponds to standard 5% damping as:

$$S_d(T)_{\xi_{eq}} = S_d(T)_{5\%}\sqrt{\frac{0.07}{0.02 + \xi_{eq}}} \qquad (2.2)$$

Note that the displacement spectrum can be calculated based on the elastic acceleration spectrum taking into account the relationship:

$$S_d(T) = \frac{S_a(T) \cdot T^2}{4\pi^2} \tag{2.3}$$

Equation (2.2) can be written in a modified form as:

$$S_d(T)_{\xi_{eq}} = S_d(T)_{5\%} c_r \tag{2.4}$$

where the coefficient c_r is expressed as:

$$c_r = \sqrt{\frac{0.07}{0.07 + 0.444\left(\frac{\mu-1}{\mu\pi}\right)}} \tag{2.5}$$

In the next step the maximum displacement Δ_u is used to estimate the corresponding period of vibration of the structure T_{sec} as it is illustrated in Fig. 2.1. Note that this period is related to the secant stiffness of the structure at the maximum displacement Δ_u.

Based on T_{sec} the corresponding stiffness k_{sec} is defined. The final step of the methods includes the estimation of the required strength of the structure. It is defined multiplying the k_{sec} by the maximum displacement Δ_u. Summarizing all the steps, the required strength of the structure can be calculated as a function of the maximum displacement Δ_u:

$$F_R = \frac{4\pi^2 m S_d^2(T_D)_{\xi_{eq}}}{T_D^2 \Delta_u} = \frac{4\pi^2 m c_r^2 S_d^2(T_D)_{5\%}}{T_D^2 \Delta_u} \tag{2.6}$$

where F_R is the required strength, m is the mass of the structure, c_r is the reduction factor depending on the ductility (see Eq. 2.5), Δ_u is the ultimate displacement, T_D is the corner period (at the end of the constant velocity region of the spectrum),

Fig. 2.1 Calculation of the effective period T_{sec} corresponding to ultimate displacement Δ_u

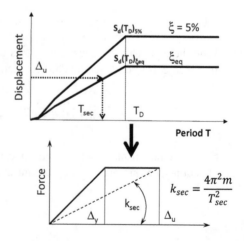

$S_d(T_D)_{\xi eq}$ and $S_d(T_D)_{5\%}$ are the corresponding spectral displacement at equivalent damping ξ_{eq} and 5% damping, respectively.

The constant yield displacement

One of the basic assumptions of the DDBD is that the yield displacement Δ_y is almost invariant once the quality of the reinforcement and the geometry of the structure are defined. For structures, which can be represented by SDOF models with reasonable accuracy, it can be estimated as:

$$\Delta_y = \frac{\phi_y(H + L_{sp})^2}{3} \qquad (2.7)$$

where Δ_y is the yield displacement, H is the effective height (in cantilever columns with the mass concentrated at the top of the column, the effective height is equal to the height of the column), L_{sp} is the strain penetration length, and ϕ_y is the yield curvature. The yield curvature can be expressed as a function of the yield deformation of steel and the height of the cross-section. For example, in the case of circular RC concrete columns the following equation has been proposed:

$$\phi_y = \frac{2.25\varepsilon_y}{D} \qquad (2.8)$$

where ε_y is the yield deformation of the steel ($\varepsilon_y = f_y/E_s$, where f_y and E_s is the yield stress and the modulus of elasticity of the steel, respectively), and D is the diameter of the cross-section. Similar expressions are proposed for other types of structural elements (columns, walls, beams, steel cross-sections), and frames (Priestley et al. 2007).

The previous observation is illustrated using the example of a cantilevered bridge column. In Fig. 2.2 the moment-curvature relationships for bridge column corresponding to different levels of the axial load is presented.

The actual moment-curvature relationship is idealized by means of a bilinear relationship, as is shown in Fig. 2.2b by a dashed line. The initial slope of this relationship defines the equivalent (effective) stiffness, which is typically defined as:

$$E_c I_{eq} = \frac{M_{y1}}{\phi_{y1}} \qquad (2.9)$$

In the above, E_c is the modulus of elasticity of the concrete, I_{eq} is the equivalent (effective) moment of inertia of the cross-section, and M_{y1} (9830 kNm) and ϕ_{y1} ($2.15 \ 10^{-3}$ 1/m) are the bending moment and the curvature corresponding to the yielding of the first layer of flexural reinforcement, respectively (see point P1 in Fig. 2.2). If there is no strain hardening, the yield moment M_y can be taken to be equal to the flexural strength M_R. The yield curvature can be estimated as $\phi_y = M_R/M_y \ \phi_{y1}$. In the presented case this value coincides quite well with the value

Fig. 2.2 The moment-curvature relationship of critical cross-section in bridge column: **a** different levels of the axial load, **b** idealization

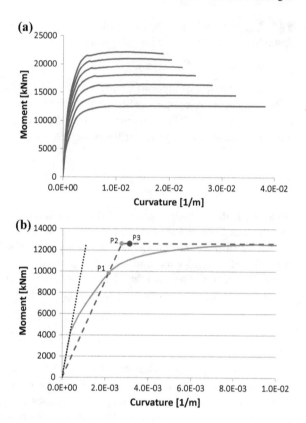

estimated by using Eq. 2.8 (see point P3 in Fig. 2.2). More details about the estimation of the yield displacement can be found elsewhere (Priestley et al. 2007).

Modified DDBD taking into account the equal displacement rule

In the original DDBD the equal displacement rule is not applied. When this rule is taken into account, the relationship between the required strength and chosen ultimate displacement Δ_u can be expressed in the same way as in the original DDBD (see Eq. 2.6); however the coefficient c_r should be replaced by:

$$c_{r1} = \frac{1}{\sqrt{\mu}} \text{ and } c_{r2} = \frac{1}{\sqrt{\frac{\mu}{1+\alpha\mu-\alpha}}} \qquad (2.10)$$

for cases with and without strain hardening, respectively (see Fig. 2.3). In these equations μ is the displacement ductility and α is the strain hardening.

The original and modified DDBD are compared in Sect. 2.2.3. It is demonstrated that in the majority of cases the original DDBD gives more conservative results than its modified version.

Fig. 2.3 Response with and without the presence of strain hardening

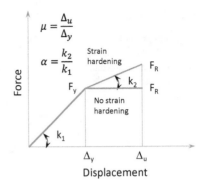

2.2.2 Basics of the SBD and the Equal Displacement Rule

Contrary to the DDBD, where the basic (or assigned) quantities are displacements and the ductility capacity of the structure, in SBD these quantities are the strength of the structure and the behaviour factor (reduction of seismic forces). Based on the strength of the structure, the ultimate displacements are typically estimated employing the equal displacement rule. The traditional interpretation of this rule (see Fig. 2.4a) can be however misleading. Thus some researchers expressed their doubts about its validity (Priestley et al. 2007). Therefore, the discussion about various interpretations of equal displacement rule is provided in sub-section *The equal displacement rule*.

Although the SBD is routinely used in the everyday design, some issues related to this type of design are discussed in sub-section *Basic equations of SBD, taking into account invariant yield displacement*. A special attention is devoted to estimation of the initial (pre-yielding) equivalent stiffness of the structure. It is defined taking into account the basic assumption of DDBD about the invariant yield displacement and taking into account the basic principle of earthquake engineering—the equal displacement rule.

The equal displacement rule

Extensive research has shown that, in the case of different systems with natural periods in the medium and long period range, the seismic demand, in terms of the displacements Δ, is independent of the strength of the system and is equal to the displacement demand Δ_e of an elastic system with the same natural period. This is the so called "equal displacement rule", which was defined by Veletsos and Newmark (1960), and has been used successfully for more than half of a century. The results of many statistical studies (e.g. Konakli and Der Kiureghian 2014) have confirmed the applicability of this rule to structures which are on firm sites with fundamental periods within the medium or long-period range, with relatively stable and full hysteretic loops.

Fig. 2.4 The equal
displacement rule:
a traditional presentation,
b an example of structures
that are used to illustrate the
equal displacement rule

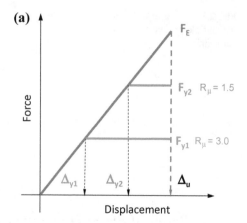

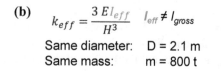

(b) $k_{eff} = \dfrac{3\,EI_{eff}}{H^3}$ $I_{eff} \neq I_{gross}$

Same diameter: D = 2.1 m
Same mass: m = 800 t

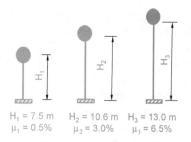

H_1 = 7.5 m H_2 = 10.6 m H_3 = 13.0 m
μ_1 = 0.5% μ_2 = 3.0% μ_3 = 6.5%

The traditional interpretation of the equal displacement rule is presented in Fig. 2.4a. It is often interpreted as the response of one (the same) structure, where different strength levels (reduction of forces) are provided. However, this interpretation is not correct, since the larger reduction of the seismic forces typically means the larger reduction of the pre-yielding stiffness, which is not the case in the Fig. 2.4a, where the pre-yielding stiffness is the same. That is why some researchers expressed their doubts about the validity of this rule.

Actually, Fig. 2.4a presents the response of three different structures, which have the same initial (pre-yielding) period of vibration (same pre-yielding stiffness and the same mass) but different strengths.

Let's say that it represents the response of three cantilever columns, presented in Fig. 2.4b, which have the same mass and diameter, but their heights and longitudinal reinforcement are substantially different. Their strengths are inversely proportional to the chosen reduction of forces (1, 1, 5, 3 for tallest, medium and shortest column, respectively). Their initial effective stiffness k_{eff} can be defined as

$$k_{eff,i} = \frac{3EI_{eff,i}}{H_i^3} \tag{2.11}$$

where E is modulus of elasticity, $I_{eff,i}$ and H_i are the effective (pre-yielding) moment of inertia and the height of the i-th column, respectively.

The yielding force of the tallest column is 3 times larger than that of the shortest column. The height of the tallest column is 1.73 times larger than that of the shortest column. The pre-yielding stiffness $k_{eff,i}$ is the same. That means that the effective moment of inertia should be proportional to the height of the columns H_i^3.

$$\frac{I_{eff,3}}{I_{eff,1}} = \left(\frac{H_3}{H_1}\right)^3 = 1.73^3 \tag{2.12}$$

Since the yielding curvature is $\phi_y = M_y/EI_{eff}$, and the ratio of the yielding moments is equal to the ratio of the effective moments of inertia ($M_{y3}/M_{y1} = 3 \cdot 1.73 = 1.73^3$) yielding curvature at the base of all columns is the same. That is compatible with the assumption that the yielding curvature depends mainly on the geometry of the cross section and the yielding deformation of the steel, which are the same in all columns.

The yielding displacement of the columns is proportional to the square of the column heights. The ratio of the yield displacements in the tallest and shortest column is:

$$\frac{\Delta_{y3}}{\Delta_{y1}} = \left(\frac{H_3}{H_1}\right)^2 = 3 \tag{2.13}$$

It is equal to the ratio of the force reduction factors. The ultimate displacement of the tallest column is $\Delta_{u3} = \Delta_{y3}$ (elastic response). The ultimate displacement of the shortest column is the same $\Delta_{u1} = 3 \Delta_{y1} = \Delta_{y3}$.

The response of the same structure with different levels of provided strength should be interpreted in different way (see Fig. 2.5). In this case the yielding

Fig. 2.5 The equal displacement rule, where correlation between the strength and the stiffness is taken into account

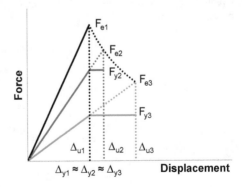

displacement is the same regardless of the level of provided strength, since the same structure (geometry) is addressed. This further means that the pre-yielding stiffness should be different (as illustrated in Fig. 2.5).

The ultimate displacements ($\Delta_{u1} - \Delta_{u3}$) are also different. However, this does not mean that the equal displacement rule is invalid. The seismic displacements Δ_{u2} and Δ_{u3} are still the same as those that characterize the corresponding elastic response, and are calculated taking into account the same effective pre-yielding stiffness (compare the dotted and solid lines of the same colour) and the mass. The ratio of the seismic displacements and yield displacements (e.g. ratio Δ_{u2}/Δ_{y2}) are still approximately the same as the corresponding level of force reduction (F_{e2}/F_{y2}). In other words, the equal displacement rule is valid, but it needs to be adequately interpreted, taking into account the correlation between the strength of the structure and the corresponding pre-yielding stiffness, as well as the corresponding reduced demand. It is applicable for each level of the chosen strength individually.

Basic equations of the SBD accounting for invariant yield displacement

Based on the discussion, presented in the previous section, it can be concluded that the pre-yielding stiffness is particularly important for SBD, since it defines the pre-yielding equivalent period of the structure, which further essentially influences the maximum displacement. Different procedures are proposed in the literature for estimation of this stiffness. In Eurocode 8/1 standard (CEN 2004) this stiffness is defined reducing the stiffness that corresponds to the gross cross-sections for 50%. This reduction can be adequate or not, depending on the level of the seismic force reduction. Following a similar procedure as the one presented in Fig. 2.1 but taking into account unreduced displacement spectrum (corresponding to 5% damping) it can be demonstrated that the pre-yielding stiffness of the structures with the periods in the constant velocity region of the spectrum (where it can be assumed that the reduction of seismic forces R_μ and the displacement ductility μ are equal; $\mu = R_\mu$) can be estimated as:

$$k_{eq} = \frac{4\pi^2 m \cdot S_d^2(T_D)}{(R_\mu \Delta_y)^2 T_D^2} = \frac{m \cdot S_a^2(T_D) T_D^2}{4\pi^2 (R_\mu \Delta_y)^2} \tag{2.14}$$

Based on this stiffness, and yield displacement (estimated according to Eq. 2.7), considering the relationship $\Delta_u = R_\mu \Delta_y$ ($\Delta_u = \mu \Delta_y$), the strength of the structure can be expressed as:

$$F_R = k_{eq} \Delta_y = \frac{4\pi^2 m c_{r1}^2 S_d^2(T_D)}{T_D^2 \Delta_u} \tag{2.15}$$

or

$$F_R = k_{eq} \Delta_y = \frac{4\pi^2 m c_{r2}^2 S_d^2(T_D)}{T_D^2 \Delta_u} \tag{2.16}$$

in the case without and with strain hardening, respectively. Coefficients c_{r1} and c_{r2} are the same as those presented in Sect. 2.2.1 (see *Modified DDBD taking into account equal displacement rule*).

2.2.3 Comparison of the SBD and the DDBD

Considering Eqs. 2.6, 2.10, 2.15 and 2.16 it is evident that the same strength is obtained by modified SBD and modified DDBD since the same basic assumptions are used: (a) equal displacement rule is applied in both methods, (b) strong correlation between the pre-yielding stiffness and the strength is taken into account, (c) yielding displacement does not depends on the strength.

In Fig. 2.6 the strength obtained in this way is compared to the values calculated using original DDBD. Different levels of ductility and three different values of the strain hardenings (0, 5% and 10%) were considered. It can be observed that the required strength, defined using original DDBD, is larger than that obtained by modified SBD and modified DDBD for all cases where the displacement ductility is larger than 2.5.

The SBD and DDBD were also compared using the numerical example of bridge column, presented in Fig. 2.7. Firstly, the required strength and the ultimate displacements were calculated using SBD, where the pre-yielding stiffness was estimated to be 50% of that corresponding to gross cross-section (traditional SBD). Than these quantities were estimated by modified SBD where pre-yielding stiffness was defined using Eq. 2.14. The yield displacement Δ_y was calculated as it was described in previous section. Finally the ultimate displacement and the corresponding strength were defined using DDBD and modified DDBD. All results are summarized in Table 2.1.

The largest strength of the structure was obtained using "traditional" SBD, since the pre-yielding stiffness was much larger than in other cases. Note that in this case the yielding moment and pre-yielding stiffness are not compatible. The pre-yielding

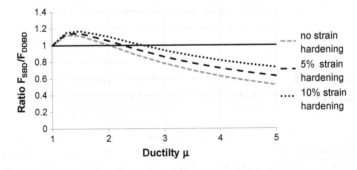

Fig. 2.6 Ratio of the strength as defined by the modified SBD (modified DDBD) and the original DDBD

Fig. 2.7 Numerical example using the SDOF representation

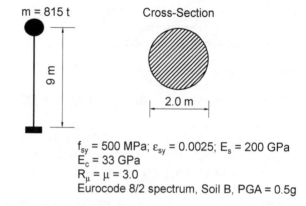

m = 815 t

Cross-Section

9 m

2.0 m

f_{sy} = 500 MPa; ε_{sy} = 0.0025; E_s = 200 GPa
E_c = 33 GPa
$R_\mu = \mu$ = 3.0
Eurocode 8/2 spectrum, Soil B, PGA = 0.5g

Table 2.1 Summary of the results, obtained using different methods

Method	F_R (kN)	D_u (cm)	D_y (cm)	I_{eff}/I_{gross}	c_r or c_{r1}
SBD	2573	14.5	4.83	0.5	–
Modified SBD	1653	22.8	7.59	0.2	0.577
DDBD	2091	22.8	7.59	–	0.653
Modified DDBD	1653	22.8	7.59	–	0.577

stiffness of 53,300 kN/m is assumed. The equivalent pre-yielding stiffness corresponding to yield moment is 21,800 kN/m. To correlate these quantities more reliably, several iterations should be performed. After certain number of iterations the equivalent pre-yielding stiffness, defined by Eq. 2.14 was obtained. The results become the same as those of the modified SBD.

According to the previous observations modified SBD and modified DDBD provided the same results. The strength calculated using original DDBD was larger for the factor $(c_r/c_{r1})^2$:

$$\frac{F_{R,SBD}}{F_{R,DDBD}} = \frac{2091}{1635} = \left(\frac{c_r}{c_{r1}}\right)^2 = \left(\frac{\sqrt{\frac{0.07}{0.07+0.444\frac{3}{3\pi}-1}}}{\frac{1}{\sqrt{3}}}\right)^2 = 1.27 \qquad (2.17)$$

2.3 An Overview of the Pushover-Based Methods Included in the Design Codes and Guidelines

The results of the design procedures, described in the previous sections, can be evaluated using the nonlinear analysis. The most accurate non-linear method is the response-history analysis. So far it is too complex for the design practice. Therefore different simplified nonlinear methods have been developed.

The most popular are different pushover-based methods, because they explicitly take into account the nonlinearity of the seismic behaviour but at the same time considerably simplify the analysis comparing to detailed response history analysis. They can be efficiently used to examine the assumptions and the response anticipated in SBD of structures.

In the majority of modern codes and design guidelines the simplest versions of pushover based methods—single-mode methods are included. Their basic properties are described very well in (Krawinkler and Seneviratna 1998):

> The static pushover analysis has no rigorous theoretical foundation. It is based on the assumption that the response of the structure (MDOF system) can be estimated using the results of the analysis of an equivalent SDOF oscillator. This means that it is assumed that the response is governed by one invariant mode of vibration. In general this is incorrect. However, the assumption is approximately fulfilled in many (regular) structures, where the influence of the higher modes is negligible and the deflection shape is almost invariable. Thus the seismic response of these MDOF systems is quite accurately estimated based on the analysis of an equivalent SDOF model.

The first step of the majority of pushover-based methods is more or less the same. The MDOF model of the structure is pushed by lateral forces (representing the inertial forces). Their intensity is gradually increased. Shear forces and displacements are monitored and correlated forming the pushover curve.

Based on the pushover curve the properties of the equivalent SDOF system of the structure are defined and the nonlinear analysis is performed. The above procedure is common to the majority of the single-mode pushover-based methods. They differ in two ways:

- regarding the procedure that is used to estimate the properties of the equivalent SDOF model
- regarding the procedure, which is used to define the response of this equivalent SDOF system.

In general there are two different approaches, which are used to define the properties of the equivalent SDOF model. They are defined either based on the equivalent pre-yielding stiffness of the structure or based on the equivalent secant stiffness corresponding to the maximum displacement (see Fig. 2.8).

The choice of the stiffness model approach typically influences also the type of the procedure, which is used to estimate the response of the SDOF model. If the SDOF model is defined based on the pre-yielding stiffness, the maximum response of the SDOF oscillator is typically estimated based on the 5% damped acceleration spectra proposed in the codes. The target displacement of the equivalent SDOF system can be estimated using the equal displacement rule (see Fig. 2.8). Since this approximation is only suitable for the medium-and long-period structures, the displacements are corrected for short-period structures. This approach is applied in, e.g., Eurocode 8/1 (EC8/1) and implicitly in FEMA-356 (2000). In FEMA-356 (2000), the maximum seismic displacements, estimated based on the analysis of SDOF system, are additionally corrected to take into account different issues which are not included in the pushover-based analysis such as strength degradation, P-Δ effect, etc.

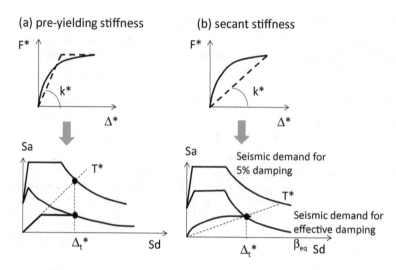

Fig. 2.8 Two different approaches, used to define properties of the equivalent SDOF system

In the second approach, where the SDOF model is defined based on the secant stiffness, the overdamped acceleration spectra are typically used (see Fig. 2.8). The capacity spectrum method approach is followed.

The application of the capacity spectrum technique means that both the structural capacity curves and the demand response spectra are plotted in the same spectral acceleration versus the spectral displacement domain and compared.

To be able to compare the capacity and demand in the same domain, the pushover curve is converted to the capacity spectrum curve using the modal shape vectors, participation factors, and modal masses obtained from a modal analysis of the structure. The capacity spectrum curve represents the relationship between accelerations S_a and displacements S_d of the equivalent SDOF oscillator. Then the standard elastic acceleration spectrum (corresponding to 5% damping) is converted to the ADRS format, where the spectral accelerations are presented as a function of the corresponding spectral displacements (see Fig. 2.8). In this way, the capacity curve and the seismic demand can be plotted on the same axes and compared.

It is assumed that the equivalent damping of the system is proportional to the area enclosed by the capacity curve. The equivalent period, T_{eq}, is assumed to be the secant period at which the seismic ground motion demands, reduced by the equivalent damping, intersect the capacity curve (FEMA-440 2005). Since the equivalent period and damping are both a function of the displacement, the solution to determine the maximum inelastic displacement (i.e., performance point) is iterative. More details about the pushover-based methods, which are included into the standards, can be found elsewhere (e.g. FEMA-440 2005; Isakovic 2014).

2.3.1 Specifics of the Analysis of Bridges

Most of the methods, which are overviewed in the previous section, have been developed primarily for the analysis of buildings. Since the response of bridges is quite different, several modifications should be introduced. They are mostly related to the construction of the pushover curve: (a) the choice of the reference location, where the displacements are monitored, (b) distribution of the lateral forces, (c) idealization of the pushover curve.

Since the detailed description of these problems and possible solutions can be found elsewhere (e.g. Kappos et al. 2012), the solutions, which are proposed at University of Ljubljana, are only summarized in this Chapter.

(a) Many standards (including Eurocode 8) propose to monitor the displacement at the centre of mass, when the pushover curve is constructed. This can be a reasonable solution as long as the location of maximum displacement coincides with the centre of mass regardless of the seismic intensity. However it should be taken into account that the location of the maximum displacement can considerably vary depending on the intensity of the seismic load. In such cases unrealistic pushover curves can be obtained if the centre of mass is chosen as a reference location. In such bridges the displacements can be considerably underestimated (see Isakovic 2014 for more details). Taking into account the extensive studies of different types of bridges (Isakovic and Fischinger 2006; Isakovic et al. 2008) performed at University of Ljubljana so far, it has been proposed to monitor the displacement at the location of the maximum displacement wherever it is.

(b) Majority of the codes suggest the use of two different distributions of the lateral load. It is recommended to take into account the envelope of the results obtained in this way. As the first option it is usually suggested to distribute the lateral load proportionally to the fundamental mode of vibration of structure.

The uniform distribution is typically suggested as the second option. In FEMA-440 (2005) it has been found that this distribution is of little value when it is used for the analysis of buildings. However, in bridges it can be useful, particularly when the higher modes have limited influence to the response (e.g. in the regions near the abutments).

In short and medium length bridges, which are supported at the abutment and which have relatively regular configuration and the response, the fundamental mode of vibration can be represented by parabolic function reasonably well. Since such distribution is easy to define, it can be used instead of the distribution proportional to the fundamental mode shape.

If the response of the bridge is governed by several modes of vibration the single-mode methods are not suitable for the analysis. The response history analysis is recommended in such cases or the multi-mode pushover based methods can be applied.

Fig. 2.9 Considerable strain
hardening can occur in
pushover curves of bridges
pinned at the abutments

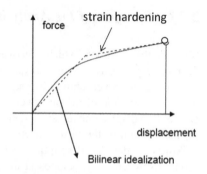

In bridges, which are not horizontally supported at the abutments, the uniform
distribution of the forces seems to be the reasonable choice. However, it should be
emphasized that the fundamental mode can be considerably different, depending on
the stiffness and the strength of columns. Thus it is recommended to use both
distributions.

In some bridges none of the distributions, described above, is the appropriate
choice, particularly in structures, which are torsionally flexible (the deck of the
bridge can heavily rotate in the horizontal plane, depending on the seismic intensity). In this type of bridges the response is often governed by one mode of
vibration, but this mode considerably changes depending on the intensity of the
seismic load. In such cases adaptive pushover-based methods should be applied.

(c) In short and medium length bridges, which are supported at the abutments, the
superstructure possesses considerable stiffness after yielding of all supporting
columns. As a consequence a considerable strain hardening can be observed in
the pushover curve (see Fig. 2.9). In such cases the bilinear idealization of the
pushover curve is recommended instead of the perfectly elasto-plastic idealization, proposed by many codes. In the case of perfectly elasto-plastic idealization, the overestimated displacements can be obtained, due to the
underestimated pre-yielding stiffness of the structure. The importance of the
pre-yielding stiffness is already discussed in Sect. 2.2.

In general it is recommended to perform iterative pushover analysis (one additional run is typically sufficient) since the pre-yielding stiffness depends on the
achieved maximum displacement. If it is considerably different than that supposed
during the idealization of the pushover curve, an additional iteration is strongly
recommended.

2.3.2 Applicability of Standard Pushover-Based Methods

The main assumption of single-mode pushover-based methods is that the response
of structure is governed by one invariant mode. Majority of these methods are
non-adaptive. That means that they suppose that the response does not considerably

change when the intensity of the seismic excitation is varied. These assumptions limit the scope of application for such methods. For example, they cannot be used for the analysis of long bridges, where the superstructure is typically quite flexible and consequently the higher modes considerably influence the seismic response.

The use of single mode methods is not recommended in bridges with very short and stiff piers, particularly if they are located close to the centre of the bridge. In such bridges higher modes are typically important, particularly for smaller intensities of the seismic load. Non-adaptive standard pushover based methods cannot be used for the analysis of torsionally sensitive structures (e.g. relatively short bridges with short and stiff central columns and with the deck, which is not supported at the abutments). More details about the applicability of these methods can be found elsewhere (e.g. Kappos et al. 2012 and Isaković 2014).

2.4 Conclusions

Two methods that can be used for the design of R/C bridges, namely the SBD and the DDBD, were analyzed in this chapter. It has been demonstrated that their results are the same when they are applied using the same set of assumptions, namely (a) the yield displacement is almost invariant for the chosen quality of the steel and the geometry of the structure, (b) the equivalent pre-yielding stiffness is strongly correlated to the strength and (c) the equal displacement rule is applied in both cases.

The strength determined by using the original DDBD, where the equal displacement rule is typically disregarded, was compared with that defined by the SBD, where the pre-yield stiffness and the strength were adequately correlated. It was found that these differences can be expressed numerically using a coefficient, which is the function of the displacement ductility μ and the type of structure. It was found that the SBD was more conservative in the case of a relatively small displacement ductility demand μ (i.e., if μ has a value of less than 2.5). In other cases, the larger required strength was obtained using the DDBD.

It has been concluded that in the SBD, the pre-yielding stiffness is directly proportional to the strength. A larger strength correlates with a larger pre-yield stiffness and vice versa. Using the typical assumption where the pre-yielding stiffness is defined to be 50% of the stiffness proportional to the gross cross-section, the equivalent stiffness as well as the required strength is often overestimated.

The equal displacement rule is also discussed. It is concluded that it should be applied for each level of the chosen strength individually, when the same structure with different strength is analyzed.

The basic assumptions and properties of the non-linear pushover-based methods, which are included into modern codes are finally reviewed. Some specifics related to their use for the analysis of bridges are presented. They are mostly related to the construction of the pushover curve as follows: (a) the choice of the reference location, where the displacements are monitored, (b) the distribution of the lateral forces and (c) the construction of the pushover curve. A short discussion on the applicability of standard pushover-based methods is also provided. It has been concluded that they can be used for the analysis of bridges where the response is governed predominantly by one mode, which is only slightly changing when the intensity of the load is varied. Typically, they cannot be used for the analysis of long bridges, where the response is highly influenced by higher modes, and for the analysis of short bridges, supported by short central columns (i.e., torsionally flexible structures), where the predominant mode of vibration changes as the intensity of the seismic excitation is varied.

References

CEN (2004) Eurocode 8: design of structures for earthquake resistance—Part 1: general rules, seismic actions and rules for buildings, EN 1998-1. European Committee for Standardization, Brussels

CEN (2005) Eurocode 8: design of structures for earthquake resistance—Part 2: bridges, EN 1998-2. European Committee for Standardization, Brussels

FEMA-356, ASCE (2000) Prestandard and commentary for the seismic rehabilitation of buildings, FEMA 356 report, prepared by the American Society of Civil Engineers for the Federal Emergency Management Agency, Washington, DC

FEMA-440, ATC, FEMA (2005) FEMA-440, improvement of nonlinear static seismic analysis procedures, applied technology council for department of Homeland Security, Federal Emergency Management Agency, Washington, DC

Isakovic T (2014) Assessment of existing structures using inelastic static analysis. Encyclopaedia of earthquake engineering. Article ID: 370159. Chapter ID: 201. doi:10.1007/978-3-642-36197-5_201-1

Isakovic T, Fischinger M (2006) Higher modes in simplified inelastic seismic analysis of single column bent viaducts. Earthq Eng Struct Dyn 35(1):95–114. doi:10.1002/eqe.535

Isakovic T, Popeyo Lazaro MN, Fischinger M (2008) Applicability of pushover methods for the seismic analysis of single-column bent viaducts. Earthq Eng Struct Dyn 37(8):1185–1202. doi:10.1002/eqe.813

Kappos AJ, Saiidi MS, Aydinoğlu MN, Isakovic T (2012) Seismic design and assessment of bridges: inelastic methods of analysis and case studies. Springer, Dordrecht

Konakli K, Der Kiureghian A (2014) Investigation of 'equal displacement' rule for bridges subjected to differential support motions. Earthq Eng Struct Dyn 43:23–39. doi:10.1002/eqe.2329

Krawinkler H, Seneviratna GDPK (1998) Pros and cons of a pushover analysis of seismic performance evaluation. Eng Struct 20(4–6):452–464

Martini S (2007) Design verification of a force- and displacement-based designed torsionally-unbalanced wall building. A dissertation submitted in partial fulfilment of the requirements for the master degree in earthquake engineering, European school for advanced studies in reduction of seismic risk, Rose School, Pavia, Italy

Priestley MJN, Calvi GM, Kowalsky MJ (2007) Displacement-based seismic design of structures. IUSS Press, Pavia

Rahman MA, Sritharan S (2011) Force-based versus displacement-based design of jointed precast prestressed wall systems, ISU-CCEE Report 01//11, A Final Report to the Precast/Prestressed Concrete, Iowa State University, Department of Civil, Construction and Environmental Engineering, Ames, USA

Veletsos AS, Newmark NM (1960) Effect of inelastic behavior on the response of simple systems to earthquake motions. In: Proceedings 2nd world conference earthquake engineering, Tokyo, Japan, vol 2, pp 895–912

Chapter 3
Multi-platform Hybrid (Experiment-Analysis) Simulations

Oh-Sung Kwon

Abstract The hybrid simulation method refers to a simulation method in which at least one substructure is experimentally tested or numerically simulated, which is integrated with a numerical model of the rest of the structural system. The integration of substructures with the rest of the system is achieved by satisfying deformation compatibility and force equilibrium at the interface of the substructures. The hybrid simulation method has been developed for seismic performance assessment of structures. In recent years, the simulation method is being further expanded to other types of loading such as temperature load due to fire. Section 3.1 of this chapter presents an overview of hybrid simulation methods. In Sect. 3.2 recent developments at the University of Toronto on hybrid simulation methods are presented.

Keywords Hybrid simulation · Multi-platform simulation · Structural performance assessment · Pseudo-dynamic simulation

3.1 Hybrid (Experiment-Analysis) Simulation Method

The first reported study of a hybrid simulation method was Hakuno et al. (1969) where a restoring force of cantilever beam was modelled physically, while an inertial mass and a damper were modelled in an analog computer. In 1975, Takanashi et al. carried out a pseudo-dynamic hybrid simulation using a digital computer. The Central Difference Method was used to numerically solve the equation of motion. Digital-to-analogue (D/A) and analogue-to-digital (A/D) converters were used for communications between a computer code written in assembly language and an actuator controller. Since then, there have been active developments on hybrid simulation methods, especially with the advancement of accessible electronic equipment (e.g. D/A, A/D converters), Internet network for

O.-S. Kwon (✉)
Department of Civil Engineering, University of Toronto, Toronto M5S 1A4, Canada
e-mail: os.kwon@utoronto.ca

© Springer International Publishing AG 2017
A.G. Sextos and G.D. Manolis (eds.), *Dynamic Response of Infrastructure to Environmentally Induced Loads*, Lecture Notes in Civil Engineering 2,
DOI 10.1007/978-3-319-56136-3_3

distributed simulation, and easily customizable actuator controllers. In the past decade, the Network for Earthquake Engineering Simulation (NEES) research program in the U.S. even further spurred the development of hybrid simulation methods as the research program's main objective was to build a network of testing facilities across the U.S. Sect. 3.1.1 presents an overview of hybrid simulation methods, which is intended to provide conceptual overview and to clarify various terminologies in hybrid simulation methods. Section 3.1.2 presents numerical integration schemes which are widely adopted in the displacement-based hybrid simulations. Section 3.1.3 presents details required for the implementation of a pseudo-dynamic hybrid simulation method. Section 3.1.4 presents two hybrid simulation examples.

3.1.1 Overview and Historical Background

The governing differential equation of a structure under dynamic load is formulated as:

$$\mathbf{m\ddot{u}} + \mathbf{c\dot{u}} + \mathbf{r}(\mathbf{u}, \dot{\mathbf{u}}) = \mathbf{f}(t) \tag{3.1}$$

where the left-hand side of the equation includes the inertial force term, $\mathbf{m\ddot{u}}$, the damping force term, $\mathbf{c\dot{u}}$, and the restoring force term, $\mathbf{r}(\mathbf{u}, \dot{\mathbf{u}})$. The damping force term is to take into account the inherent energy dissipation in the structure that is difficult to model explicitly. In a linear elastic system the restoring force term, $\mathbf{r}(\mathbf{u}, \dot{\mathbf{u}})$, is a linear function of stiffness and displacement, i.e. $\mathbf{ku}$. In an inelastic system, the restoring force term can be a nonlinear function of displacement, $\mathbf{u}$, or velocity, $\dot{\mathbf{u}}$. The right-hand side of the equation, $\mathbf{f}(t)$, is a dynamic force term which varies over time, t. When the response of an entire structural system is evaluated numerically, the above equation can be solved using various numerical time integration schemes.

In a hybrid simulation, all or part of the restoring force term, $\mathbf{r}(\mathbf{u}, \dot{\mathbf{u}})$, are evaluated experimentally. Thus, the restoring force term can be separated into two components:

$$\mathbf{r} = \mathbf{r}_E + \mathbf{r}_N \tag{3.2}$$

where $\mathbf{r}_E$ and $\mathbf{r}_N$ are force vectors evaluated from an experimental specimen and a numerical model, respectively. Both terms might be a function of displacement, $\mathbf{u}$, and/or velocity, $\dot{\mathbf{u}}$. The terms in the parenthesis in Eq. (3.1), $(\mathbf{u}, \dot{\mathbf{u}})$, are removed for brevity of the expression. If the experimentally evaluated restoring force term, $\mathbf{r}_E$, depends on the rate of deformation (i.e. velocity), then the experimental specimen should be tested at the actual rate of loading. Depending on how the Eq. (3.2) is evaluated, a hybrid simulation can be classified in several different ways as presented below.

3.1.1.1 Pseudo-dynamic and Real-Time Hybrid Simulations

A pseudo-dynamic (PsD) simulation generally refers to a hybrid simulation method whether the displacement is imposed to a physical specimen with an expanded time scale. This type of simulation method is ideal when the hysteretic behaviour of the physically tested specimen does not depend on the rate of loading. For example, the restoring force of steel or concrete does not largely depend on the strain rate that may be caused by earthquake events. In such case, a PsD simulation is a suitable simulation method to evaluate the dynamic response of a structure. Because the experiment is carried out at an extended time scale, this simulation method is beneficial in several ways. For example, it is possible to acquire and review detailed experimental data during experiment, which gives an opportunity to take a corrective measure if any unforeseen issues arise during an experiment. In addition, the slow-rate of experiment allows the use of sophisticated inelastic numerical model for the rest of the structural system such that both the experimental specimen and the numerical model can be rigorously represented. When a stiff specimen is tested, such as reinforced concrete columns, the reaction system of the specimen develops non-negligible elastic deformation. The pseudo-dynamic simulation allows iterations within an integration time step to correct the displacement error resulting from the elastic deformation. On the other hand, the pseudo-dynamic simulation cannot be used for a specimen with rate-dependent characteristics such as viscous dampers, tuned liquid dampers, magneto-rheological dampers, etc. Several frameworks have been developed for PsD simulations such as UI-SimCor (Kwon et al. 2005, 2008), P2P internet online hybrid system (Pan et al. 2006), and OpenFresco (Schellenberg et al. 2008, 2009).

In the real-time hybrid simulation, the displacement is imposed to a physical specimen at the actual rate of loading. This simulation method is ideal to evaluate the performance of a structural system with rate-dependent structural elements. In real-time hybrid simulations, there are two main challenges which have been investigated extensively in the past decade: real-time analysis of a numerical model and compensation of actuator delay.

In most real-time simulations, the numerical model needs to be greatly simplified as a linear elastic system or with very limited nonlinearity with limited number of degrees of freedoms. Because the simulation needs to be carried out at the actual rate of loading, it is challenging to use a sophisticated numerical model, which often cannot complete one step of analysis within the actual time step. Several developments have been made to improve the computing speed, and to use more realistic numerical models. For example, Karavasilis et al. (2008) developed a non-linear two-dimensional frame analysis program, HybridFEM, which can run on MATLAB/Simulink environment which is compatible with Target PC for real-time hybrid simulation. Saouma et al. (2012) developed a finite element package, Mercury, which is written both in C++ computer language and MATLAB. The program can be compiled and embedded in MATLAB/Simulink or LabVIEW code for real-time execution. In both studies, the analysis software needed to be optimized to minimize computing time.

Another main challenge in real-time hybrid simulations is the delay (or time lag) in the control of actuators. The delay in the execution of command by the actuator has an effect of negative damping, which adds energy to a structural system. The negative damping effect may lead to unstable response of the structure due to the additional energy from the actuator. Many different approaches have been proposed to compensate the delay in real-time hybrid simulation (Ahmadizadeh et al. 2008; Carrion and Spencer 2007; Chae et al. 2013; Chen and Ricles 2009; Darby et al. 2002; Horiuchi et al. 1999; Nakashima and Masaoka 1999; among many others).

Real-time hybrid simulation: displacement and effective force-based simulations

Most of the pseudo-dynamic or real-time hybrid simulations are carried out based on predicted displacements from a time integration scheme. Once the predicted displacements are imposed to physical specimens or numerical elements, then a restoring force vector is assembled. In the displacement-based hybrid simulations, the inertial, damping, and external forces are represented in a numerical model. Thus, it is essential to have a realistic representation of the initial stiffness, damping, and mass matrices which are used to predict the next step's displacements. In some experimental specimens, especially when there exists multiple coupled DOFs, evaluation of the stiffness matrix of an experimental specimen is not a trivial task. In addition, when a specimen is very stiff, such as a simulation of reinforced concrete column subjected to vertical excitation, it is difficult to control the deformation of the specimen up to the precision required for a hybrid simulation.

The effective force testing method was developed to overcome the above limitations. In the effective force testing method, the effective force, right-hand side of Eq. (3.1), $\mathbf{f}(t) = -\mathbf{m}l\ddot{u}_g$, is directly applied to a specimen as a pre-defined force history. Thus, the experimental specimen should include realistic inertial mass. Because the inertial force and other rate-dependent characteristics of specimens are physically modelled, the effective force testing needs to be carried out in real time. Similarly to the real-time hybrid simulation, the effective force testing method requires accurate control of actuators. Examples of effective force testing include (Chen 2007; Dimig et al. 1999; Mahin and Shing 1985; Zhao et al. 2006).

3.1.1.2 Conventional or Sub-structure Hybrid Simulation

In hybrid simulations an entire structural system can be modelled experimentally while the inertial mass, damping, and external force terms are modelled numerically. In this approach, there is no restoring force calculated from a numerical model, i.e. $\mathbf{r} = \mathbf{r}_E$ in Eq. (3.2). This simulation approach is feasible only when a testing facility can accommodate the entire structural system, and only when there is a sufficient number of testing equipment (i.e. servo-controlled actuators). This simulation method is more accurate than the other approach (sub-structure hybrid simulation) because the restoring forces of the entire system are evaluated experimentally without many simplifying assumptions in a numerical model. Examples of this type of hybrid simulations are Chae et al. (2014) and Negro et al. (1996). On the other hand, this approach requires significant experimental resources such as space, equipment, technician time, fund, etc. to construct a large scale experiment. Considering the scale of a typical structural testing facility, three- to four-storey building is probably the largest scale of a structure that can be tested as a whole.

In substructure hybrid simulations, the restoring force term, Eq. (3.2), is evaluated as a summation of restoring force vectors from more than one substructures, i.e. one or more experimental substructures, and one or more numerical substructures. Structural elements that cannot be accurately represented using numerical models are represented with physical specimens. Since most of the structural systems are numerically modelled, the sub-structure hybrid simulation method allows simulation of a large scale structure. This simulation method, however, improves the accuracy of the simulation only when there is sufficient interaction between the physically tested elements and the rest of the structural system. Examples of sub-structure hybrid simulations are Dermitzakis and Mahin (1985), Elnashai et al. (2008), Kammula et al. (2014), Murray et al. (2015), Mahmoud et al. (2013) and Spencer et al. (2006).

3.1.1.3   On-Site or Distributed Hybrid Simulation

Hybrid simulation can be carried out in a single site, i.e. on-site hybrid simulation. If the restoring force vector in Eq. (3.2) can be assembled through an Ethernet or the Internet network, it is possible to run a hybrid simulation at a multiple geographically distributed sites. For example, in the MISST project (Spencer et al. 2006), one of the bridge pier was tested at Lehigh University while the other pier was tested at the University of Illinois at Urbana-Champaign. These two physical specimens were integrated with a numerical model of the rest of the bridge. This approach, in theory, can take advantage of many unique testing facilities around the world. Yet, implementation and coordination of the distributed hybrid simulation is not a trivial task. Several distributed hybrid simulations have been reported in literature (Mosqueda et al. 2004, 2008a, b; Sextos et al. 2014). Up to now, most of the distributed hybrid simulations were to develop and validate simulation methodology rather than to apply the simulation method to take advantage of distributed testing facilities and to develop new knowledge from the tested specimen.

3.1.1.4   Step-Wise or Continuous Hybrid Simulation

Pseudo-dynamic hybrid simulation is carried out by imposing displacement steps. The displacement steps are imposed as smooth harmonic ramp function, which is followed by a hold period. This ramp and hold type experiment allows accurately imposing displacement to a specimen and accurately measuring deformation and forces. By introducing a hold period, the issues related to the time lag in digital

filters can be avoided. In real-time hybrid simulation, the displacement commands
are updated at a fast rate (e.g. 1024 Hz). Thus, the real-time hybrid simulation is
inherently continuous testing method. The control method for real-time hybrid
simulation can be used to run a pseudo-dynamic hybrid simulation at an expanded
time scale, which then becomes continuous pseudo-dynamic hybrid simulation.
This type of simulation method is beneficial to avoid stress relaxation in some
tested materials during the hold period. Examples of continuous hybrid simulations
are Magonette (2001), Mosqueda et al. (2004), Nakashima et al. (1993), Takanashi
and Ohi (1983) and Watanabe et al. (2001).

3.1.1.5 Multi-platform Hybrid Simulation

The framework for hybrid simulation can be extended to substructure numerical
simulations. In the research on conventional substructure analysis, the primary
focus has been on reducing the computing time by substructuring and distributing
the computational demand to multiple processors. This type of substructure analysis
was typically carried out using a single software in a computer cluster. In hybrid
simulation method, however, two distinctively different substructures can be integrated; the substructures can be either experimental specimens or numerical models.
Thus, by using a hybrid simulation framework, it is possible to integrate one or
more numerical models that are analyzed in different software. This approach,
which is referred to as a multi-platform simulation, allows integration of dedicated
software in diverse problem domains, such as soil-foundation system, reinforced
concrete structures, etc., to simulate a large and complex structural system.
UI-SimCor (Kwon et al. 2005, 2008) and OpenFresco (Schellenberg et al. 2008)
has capability to extend the framework for multi-platform simulation. Recently,
Huang and Kwon (2015) proposed a standardized data exchange format and protocol to easily integrate diverse numerical analysis software and experimental
specimens. Some of the examples of multi-platform simulation are Kwon and
Elnashai (2008) and Sadeghian et al. (2015).

3.1.2 Numerical Integration Schemes

Numerical methods to solve the Eq. (3.1) can be categorized as explicit and implicit
methods. In the explicit method, step $n + 1$'s displacement and velocity are calculated based on the displacement, velocity, and acceleration of step n, and the
external force of step n and $n + 1$. Because the external force history is assumed to
be known a priori, all values are known to calculate step $n + 1$'s displacement and
velocity even without using Eq. (3.1) at step $n + 1$. Because the response values of
the step $n + 1$ is calculated based on previous step's response values without using
Eq. (3.1) at step $n + 1$, explicit schemes do not fully satisfy equilibrium at each
time step. In addition, explicit schemes tend to have stability issue when the time

step is greater than a threshold value. Thus, explicit scheme requires small time step in comparison with the shortest natural period of a structural system.

In the implicit method, the force equilibrium is satisfied by finding a solution of the Eq. (3.1) at each time step. Thus, implicit method requires iteration in an inelastic system, which poses difficulty in the application of an implicit method in hybrid simulations. There has been many algorithmic developments on numerical time integration schemes for hybrid simulations in several ways such as; improving the stability and accuracy of explicit time integration scheme (Chang 2010; Chen and Ricles 2008); avoiding iterations in physical substructures in an implicit scheme (Bursi et al. 2010; Combescure and Pegon 1997; Mosqueda and Ahmadizadeh 2011; Nakashima et al. 1993); using iterative implicit scheme (Shing et al. 1991).

In the following, the widely used time integration scheme (α-OS method) in hybrid simulations is summarized to illustrate the overall procedure of running time integration scheme in a hybrid simulation. Newmark's integration scheme (Newmark 1959) is first introduced which forms the basis of the α-operator splitting (α-OS) time integration scheme (Combescure and Pegon 1997; Nakashima et al. 1993). It is suggested to refer latest literature listed in the previous paragraph for newly developed algorithms.

3.1.2.1 Newmark's Integration Scheme

In the numerical solution of the equation of motion, Eq. (3.1), the solution is evaluated at each discrete time step, typically at a uniform interval, Δt. The time, t_n, refers to nth multiple of Δt, i.e. $t_n = n\Delta t$. The variables $\ddot{\mathbf{u}}_n$, $\dot{\mathbf{u}}_n$, $\mathbf{u}_n$, and $\mathbf{f}_n$, denote acceleration, velocity, displacement, and external force at t_n. These values at discrete time should satisfy the dynamic equilibrium equation, Eq. (3.1).

$$\mathbf{m}\ddot{\mathbf{u}}_n + \mathbf{c}\dot{\mathbf{u}}_n + \mathbf{r}_n = \mathbf{f}_n \qquad (3.3)$$

Once the relationships between the responses at t_{n+1} and t_n are established, the responses at any discrete time step i can be found by recursively applying the relationships from t_0 to the time step i. In the well-known Newmark's family of time-stepping method, the relationships between the responses at t_{n+1} and t_n are defined as below.

$$\dot{\mathbf{u}}_{n+1} = \dot{\mathbf{u}}_n + [(1-\gamma)\Delta t]\ddot{\mathbf{u}}_n + (\gamma\,\Delta t)\ddot{\mathbf{u}}_{n+1} \qquad (3.4)$$

$$\mathbf{u}_{n+1} = \mathbf{u}_n + \Delta t\,\dot{\mathbf{u}}_n + [(0.5-\beta)\,\Delta t^2]\ddot{\mathbf{u}}_n + (\beta\,\Delta t^2)\ddot{\mathbf{u}}_{n+1} \qquad (3.5)$$

where the parameters, γ and β, define how the acceleration varies from step n to step $n+1$. For example, $\gamma = 1/2$ and $\beta = 1/4$ is for the case when acceleration is constant between the two time steps. $\gamma = 1/2$ and $\beta = 1/6$ is analogous to the linear variation of acceleration between steps n and $n + 1$. Selection of these parameters also affect the stability and accuracy characteristics of the integration scheme.

Note that the Eqs. (3.4) and (3.5) includes an unknown term $\ddot{\mathbf{u}}_{n+1}$ which needs to be found using the equilibrium equation at step $n + 1$,

$$\mathbf{m}\ddot{\mathbf{u}}_{n+1} + \mathbf{c}\dot{\mathbf{u}}_{n+1} + \mathbf{r}_{n+1} = \mathbf{f}_{n+1} \tag{3.6}$$

For a linear elastic system, the restoring force, $\mathbf{r}_{n+1}$, can be evaluated using a stiffness matrix, Eq. (3.7), and the responses at step $n + 1$ can be found without iterations by substituting Eqs. (3.4), (3.5), and (3.7) into Eq. (3.6).

$$\mathbf{r}_{n+1} = \mathbf{k}\mathbf{u}_{n+1} \tag{3.7}$$

For a nonlinear system, however, the above set of equations need to be solved iteratively using a nonlinear solution method such as Newton-Raphson procedure. Because the responses at step $n + 1$ are found by enforcing equilibrium at step $n + 1$, the Newmark's method is an implicit scheme.

3.1.2.2 α-Operator Splitting (α-OS) Method

In the operator splitting method, the Eqs. (3.4) and (3.5) are split into known terms from step n and unknown terms at step $n + 1$.

$$\dot{\mathbf{u}}_{n+1} = \tilde{\dot{\mathbf{u}}}_{n+1} + (\gamma\Delta t)\ddot{\mathbf{u}}_{n+1} \tag{3.8}$$

$$\mathbf{u}_{n+1} = \tilde{\mathbf{u}}_{n+1} + (\beta\,\Delta t^2)\ddot{\mathbf{u}}_{n+1} \tag{3.9}$$

where

$$\tilde{\dot{\mathbf{u}}}_{n+1} = \dot{\mathbf{u}}_n + [(1-\gamma)\Delta t]\ddot{\mathbf{u}}_n \tag{3.10}$$

$$\tilde{\mathbf{u}}_{n+1} = \mathbf{u}_n + \Delta t\,\dot{\mathbf{u}}_n + [(0.5-\beta)\,\Delta t^2]\ddot{\mathbf{u}}_n \tag{3.11}$$

Equations (3.10) and (3.11) are referred to as predicted displacement and velocity, and Eqs. (3.8) and (3.9) are corrected values. In the operation splitting method, the restoring force term in Eq. (3.6) is approximated as a summation of restoring force due to the predicted displacement (Eq. 3.10) and velocity (Eq. 3.11), and the force proportional to the initial stiffness, $\mathbf{k}^1$, and the difference between the corrected and the predicted displacements:

$$\begin{aligned}\mathbf{r}_{n+1} &\approx \tilde{\mathbf{r}}_{n+1} + \mathbf{k}^1(\mathbf{u}_{n+1} - \tilde{\mathbf{u}}_{n+1})\\ &= \tilde{\mathbf{r}}_{n+1} + \mathbf{k}^1(\beta\,\Delta t^2)\ddot{\mathbf{u}}_{n+1}\end{aligned} \tag{3.12}$$

In the operator-splitting method, the Eqs. (3.8) and (3.12) are substituted to Eq. (3.6) to find the unknown acceleration at step $n + 1$, $\ddot{\mathbf{u}}_{n+1}$, which can be subsequently used to correct the displacement and the velocity using Eqs. (3.8) and (3.9).

As it can be found from the above steps, the operator splitting method uses the explicit method to predict displacement and velocities (Eqs. 3.10 and 3.11) with which the restoring force from a structure is evaluated. Thus, this method is called *non-linearly explicit*. If the system is linear elastic, however, the Eq. (3.12) becomes $\mathbf{r}_{n+1} = \mathbf{k}^{\mathrm{I}}\mathbf{u}_{n+1}$, which results in the implicit scheme. Thus, the operator-splitting method is *linearly implicit* method.

The operator splitting method can be used with the α-modified Newmark scheme to suppress undesired oscillation, which is referred to as α-OS method. The modification is made to the equilibrium equation, Eq. (3.8), by introducing a weighting factor, α, between the values at steps n and $n + 1$.

$$\mathbf{m}\ddot{\mathbf{u}}_{n+1} + (1+\alpha)\mathbf{c}\dot{\mathbf{u}}_{n+1} - \alpha\mathbf{c}\dot{\mathbf{u}}_n + (1+\alpha)\mathbf{r}_{n+1} - \alpha\mathbf{r}_n = (1+\alpha)\mathbf{f}_{n+1} - \alpha\mathbf{f}_n \quad (3.13)$$

where $\alpha \in [-1/3, 0]$ In the α-OS method, the parameters β and γ are chosen as:

$$\beta = (1-\alpha)^2/4, \quad \gamma = (1-2\alpha)/2 \quad (3.14)$$

Once Eqs. (3.8) and (3.12) are substituted to Eq. (3.13), the acceleration at step $n + 1$ can be found as below.

$$\hat{\mathbf{m}}\ddot{\mathbf{u}}_{n+1} = \hat{\mathbf{f}}_{n+1} \quad (3.15)$$

where the equivalent mass matrix, $\hat{\mathbf{m}}$, and the equivalent force vector, $\hat{\mathbf{f}}_{n+1}$, are:

$$\hat{\mathbf{m}} = \mathbf{m} + \gamma\,\Delta t(1+\alpha)\mathbf{c} + \beta\,\Delta t^2(1+\alpha)\mathbf{k}^{\mathrm{I}} \quad (3.16)$$

$$\begin{aligned}\hat{\mathbf{f}}_{n+1} = (1+\alpha)\mathbf{f}_{n+1} - \alpha\mathbf{f}_n + \alpha\tilde{\mathbf{r}}_n - (1+\alpha)\tilde{\mathbf{r}}_{n+1} + \alpha\mathbf{c}\tilde{\mathbf{u}}_n \\ - (1+\alpha)\mathbf{c}\tilde{\mathbf{u}}_{n+1} + \alpha(\gamma\,\Delta t\,\mathbf{c} + \beta\,\Delta t^2\,\mathbf{k}^{\mathrm{I}})\ddot{\mathbf{u}}_n\end{aligned} \quad (3.17)$$

The overall step-by-step procedure to implement the α-OS method in hybrid simulations is summarized in Fig. 3.1.

Combescure and Pegon (1997) extensively investigated the numerical properties of the α-OS method. The study found that the numerical damping introduced by the α parameter can limit the impact of spurious high-frequency oscillation that results from experimental errors. In addition, the I-modification, which compensates experimental error in imposing the predicted displacement (Step 7 in Fig. 3.1), can limit the impact of undershoot error.

1. Choose Δt. Evaluate equivalent mass, $\hat{\mathbf{m}}$, using Eq. (3.16).
2. Set n=0. Initialize $\tilde{\mathbf{u}}_0, \mathbf{u}_0 = \tilde{\mathbf{u}}_0, \dot{\mathbf{u}}_0, \ddot{\mathbf{u}}_0, \tilde{\mathbf{r}}_0$ and $\mathbf{f}_0$
3. Calculate external force at step n+1, $\mathbf{f}_{n+1}$
4. Calculate predicted displacement, $\tilde{\mathbf{u}}_{n+1}$, and velocity, $\dot{\tilde{\mathbf{u}}}_{n+1}$, using Eq. (3.11) and (3.10), respectively.
5. Impose $\tilde{\mathbf{u}}_{n+1}$ to substructures (physical specimen and numerical substructure)
6. Measure restoring force, $\tilde{\mathbf{r}}_{n+1}^m$, and imposed displacement, $\tilde{\mathbf{u}}_{n+1}^m$
7. Calculate corrected restoring force $\tilde{\mathbf{r}}_{n+1} = \tilde{\mathbf{r}}_{n+1}^m - \mathbf{k}^{\mathrm{I}}(\tilde{\mathbf{u}}_{n+1}^m - \tilde{\mathbf{u}}_{n+1})$
8. Calculate equivalent force vector $\hat{\mathbf{f}}_{n+1}$ using Eq. (3.17)
9. Solve linear equation Eq. (3.15) to find $\ddot{\mathbf{u}}_{n+1}$
10. Calculate corrected displacement, $\mathbf{u}_{n+1}$, and velocity, $\dot{\mathbf{u}}_{n+1}$, using Eqs. (3.8) and (3.9)
11. Set $n = n+1$ and go to Step 3.

Fig. 3.1 Implementation of α-OS method for hybrid simulation (Combescure and Pegon 1997)

3.1.3 Implementation of Pseudo-dynamic Hybrid Simulation

Hybrid simulation methods have been developed for more than four decades as presented in Sect. 3.1.1. Yet, it still requires substantial amount of time and efforts to implement and run a hybrid simulation for the first time. This section discusses a few details related to the implementation of a hybrid simulation, specifically a PsD hybrid simulation. There are mainly three components that need to be implemented to run a PsD hybrid simulation: an integration module, substructure module(s), and a communication mechanism as illustrated in Fig. 3.2.

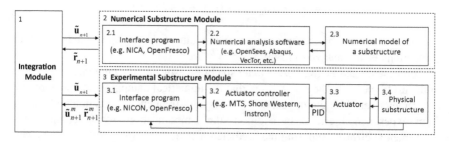

Fig. 3.2 Typical architecture of hybrid simulation

3.1.3.1 Integration Module

An integration module, Block #1 in Fig. 3.2, is a main software module which runs a numerical time integration scheme. In most hybrid simulation frameworks, such as OpenSees (Mazzoni et al. 2007), HybridFEM (Karavasilis et al. 2008), and Mercury (Saouma et al. 2012), one computer program is used to run a numerical time integration scheme, which also includes finite elements of the majority of a structural model. In UI-SimCor (Kwon et al. 2008), on the other hand, only a numerical time integration scheme is implemented. All restoring forces are assembled from substructure modules (i.e. numerical substructure or physical specimens). The numerical time integration scheme in the integration modules predicts target displacements and/or velocities, which are imposed to substructure modules. In a purely numerical analysis of an inelastic system, several nonlinear solution scheme can be tried to find a solution. In this process, a nonlinear solution scheme can impose 'trial' displacement and later cancel the displacement if a converged solution cannot be found. In a hybrid simulation with physical specimens, however, it is not possible to impose a trial displacement and cancel the displacement later even if there is a numerical convergence issue. Thus, time integration schemes, which do not require iterations within a time step, is most widely used in PsD hybrid simulations. Any structural analysis software with time integration schemes without iteration can be used as an integration module as long as the software allows implementation of a new element which can communicate with an external program (i.e. substructure module) through a computer network.

3.1.3.2 Substructure Module

A substructure module, Blocks #2 and #3 in Fig. 3.2, refers to a module in which a numerical substructure is analysed or a physical substructure is tested. The substructure modules impose the predicted displacements received from the integration module, and return the calculated or measured restoring forces back to the integration module. In most of the substructure multi-platform simulations, the substructure modules do not include inertial masses, and are assumed to be either displacement or velocity-dependent.

To impose displacement or velocity to a numerical model at each time step, it is necessary to keep the state of the numerical model at the previous time step, and impose the newly predicted displacement (or velocity) to the numerical model. Most numerical analysis tools require an access to source code or a user-customizable elements to implement this feature, i.e. changing the boundary condition on the fly without restarting the program. The hybrid simulation frameworks, UI-SimCor and OpenFresco, provides interface (Block #2.1 in Fig. 3.2), between a numerical integration scheme and a few analysis packages (Block #2.2 in Fig. 3.2) such as Abaqus (Simulia 2014), OpenSees (Mazzoni et al. 2007), VecTor program suite (Vecchio and Wong 2003), etc.

To impose a displacement to a physical specimen, it is necessary to input the displacement as a target command to an actuator controller (Block #3.2 in Fig. 3.2), which typically runs PID control loop. Mainly two methods have been used to impose the displacement command to an actuator controller; analogue voltage signals or shared memory approach. Based on the experience of the author of this Chapter, most actuator controllers from MTS, Shore Western, and Instron, can run the PID control loop based upon the external command from analogue voltage signals. The communication with analogue voltage signals require D/A and A/D converters to convert digital values (i.e. predicted displacements) to voltage signals, and to convert measured responses (i.e. displacements and forces) back to digital values, which are returned to an integration module. The cost of the D/A and A/D converters depend on the number of channels, resolution, sampling rate, etc. In general, these equipment are relatively affordable in comparison with the shared memory approach. The shared memory approach requires direct access to a memory block that is used as the source of command in an actuator controller. This approach requires a proprietary hardware, such as SCRAMNet cards, to allow multiple computers access same memory block.

3.1.3.3 Communication Mechanism

Figure 3.2 presents an overview of communication layers in a hybrid simulation. Depending on whether the simulation is a real-time or a PsD hybrid simulation, some of the blocks in the figure can be merged to minimize communication needs between blocks. Using an Ethernet or the Internet for the communication between the integration module (Block #1) and substructure modules (Block #2 and #3) are preferable especially when diverse numerical substructure modules are planning to be used, or when a geographically distributed simulation is planned. The communication through the Internet, however, is subjected to inherent latency (time lag) and jitter (variability of time lag) which depends on physical routes of the communication and network traffic. Thus, at this stage, distributed real-time simulation through the Internet is still very challenging. In real-time hybrid simulations, the communication between the integration and substructure modules is established either through closed Ethernet network or function calls, which guarantee deterministic delivery of messages between modules.

3.1.3.4 Interface Program for Actuator Controllers

When a physical substructure element is tested, it is necessary to impose displacements either through analogue voltage signals or a shared memory block as discussed in the above. Thus, it is typical to use an interface program, Block 3.1. The interface program needs to be equipped with several other functionalities as summarized below. The following discussion is primarily based on the functionalities implemented in the Network Interface for Controllers (NICON, Zhan and

Kwon 2015) that has been developed at the University of Toronto and used in several hybrid simulations.

Transformation of coordinate system

The integration module always run time integration scheme based on the global coordinate system. Yet, the coordinates of the actuators in a testing facility may not be aligned with the global coordinate system. In addition, imposing displacements at a control point by changing the stroke of actuators require nonlinear transformation between the displacement commands in the global coordinate system to the actuators' strokes. The transformation of coordinate system requires several steps as illustrated in Fig. 4.3. A frame element's displacement in a 3D space (Fig. 3.3a) needs to be converted to the element's coordinate system (Fig. 3.3b). Because an element develops restoring force only due to a relative deformation unless inertial effect is considered, the rigid body component of the deformation needs to be removed (Fig. 3.3c). Then, the element's relative deformation needs to be aligned with the testing setup's Cartesian coordinate system (Fig. 3.3d) which, in turn, needs to be transformed to the actuators' stroke (Fig. 3.3e). Once the actual strokes from the internal LVDTs and the forces from load cells are measured in the actuators' coordinate system, all the measured values need to be transformed back to the global coordinate system (Fig. 3.3a).

In addition, depending on the imposed boundary conditions, the experimental setup may have diverse configurations of the actuators. Most typical testing configurations in the field of civil engineering are summarized in Fig. 3.4. There might be several other variations in addition to these configurations. Thus, unless the implementation of a hybrid simulation is only for one specific experiment, the interface program (Block #3.1) needs to generalized to accommodate various actuator configurations.

Error compensation

In many cases, the reaction system in a structural test is not rigid enough. The compliance of the reaction system does not significantly influence the test results in quasi-static cyclic tests. In hybrid simulations, the response of the specimen at each

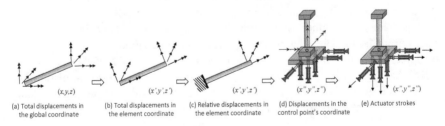

(a) Total displacements in the global coordinate   (b) Total displacements in the element coordinate   (c) Relative displacements in the element coordinate   (d) Displacements in the control point's coordinate   (e) Actuator strokes

Fig. 3.3 Conversion of coordinate system for hybrid simulation (after Zhan and Kwon 2015)

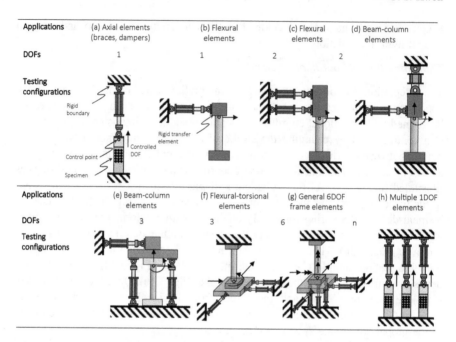

Fig. 3.4 Configuration of actuators for hybrid simulations (after Zhan and Kwon 2015)

time step influence the response of the structural system. Thus, if there is a systematic error at each time step such as compliance of the reaction system, the error accumulates over many time steps which lead to inaccurate seismic response of the structural system. To compensate the error resulting from elastic deformation, it is necessary to measure actual deformation of the specimen in addition to the measurement from the stroke of the actuators. There are mainly two methods to compensate this error. In the first method, the PID control loop in Block 3.2 can be run based on the actual deformation of the specimen (Block 3.4) rather than the stroke of the controlled actuators (Block 3.3). This approach, however, can be quite dangerous if the specimen develops large unforeseen failure modes. The second approach is using an outer loop where the interface program (Block 3.1) can correct the error based on the actual deformation of the specimen (Block 3.4).

There are several other critical functionalities that are required in the interface program including data logging for debugging purpose, limit checks for safety of equipment and specimen, etc. The interface program, NICON (Zhan and Kwon 2015), has been developed to make it easier to implement a hybrid simulation in a testing facility which is new to a hybrid simulation method. The program includes the essential features discussed in the above.

3.1.4 Hybrid Simulation Examples

This section presents two examples of hybrid simulations. The first example is a geographically distributed hybrid simulation of a bridge where numerical substructure models and one experimental specimen were distributed among different countries in North America and in Europe (Sextos et al. 2014; Bousias et al. 2014). The second example is a hybrid simulation of a six-storey building with Self-Centering Energy Dissipating (SCED) braces (Kammula et al. 2014).

3.1.4.1 Inter-continental Distributed Hybrid Simulation of a Three Span Bridge

Studied on distributed hybrid simulations has been reported since mid-2000s (Elnashai et al. 2008; Kim et al. 2012; Mosqueda et al. 2008a, b; Ojaghi et al. 2014; Pan et al. 2005; Spencer et al. 2004; Cortes-Delgado and Mosqueda 2010; Takahashi and Fenves 2006; Whyte et al. 2010). In a distributed hybrid simulation, the latency and jitter of the Internet network are dominant factors that influence the rate of the simulation. For example, in a typical single-site hybrid simulation, the delay resulting from the actuation and filtering of signal is in the order of 10–15 ms. In comparison, the average round-trip time of data packets between University of Toronto, Canada and the Aristotle University of Thessaloniki, Greece varies between 170–220 ms depending on the time of testing (Maynard and Kwon 2014). This level of latency is very hard to overcome if one wants to run a distributed real-time hybrid simulation. As shown in Fig. 4.6, the range of latency from consecutive round-travel time tests can be from 173 to 215 ms. On the other hand, at a closer distance the latency is smaller, which might be overcome through a delay compensation scheme; for example 7–8 ms between the University of Toronto and the University of Illinois at Urbana-Champaign (Maynard and Kwon 2014) and 50–60 ms between the University of Oxford and the University of Bristol (Ojaghi et al. 2014). The latter includes the delay that result from application of commands and measurement of data. The issues with the latency and delay becomes more complicated if multiple sites at various locations are involved (Fig. 3.5).

Thus, the inter-continental distributed hybrid simulations were carried out to investigate the effect of remote host distance on the feasibility of executing hybrid simulation among distant sites and to demonstrate the stability of an intercontinental multi-platform and/or hybrid simulation.

A reference structure that was evaluated through the hybrid simulation is a three-span bridge overcrossing a highway as shown in Fig. 3.6. The bridge deck is supported on elastomeric bearings at the abutments. The bridge was sub-structured into five modules; two bridge piers, two elastomeric bearings, and the deck. One of the bridge bearings (Module 4 in Fig. 3.6) was represented with a physical specimen at the University of Patras. All other substructure modules were represented with numerical models in OpenSees. UI-SimCor (Kwon et al. 2008) was used as a

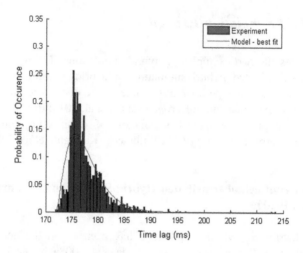

Fig. 3.5 Histogram of round-travel time packet test between the University of Toronto and the Aristotle University of Thessaloniki (Maynard and Kwon 2014)

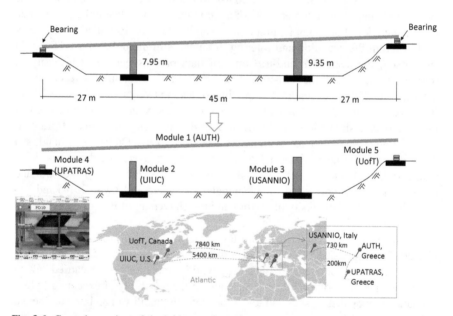

Fig. 3.6 General overview of the bridge configuration

main integrator for the hybrid simulation. Considering the involvement of multiple sites and the network latency, the experiment was carried out as a pseudo-dynamic simulation. The rate dependency of the physical specimen was compensated for by applying a scale factor in the measured force. An independent component test confirmed that such compensation can reproduce the behaviour of the bearing test at the actual rate of loading.

From the hybrid simulation, it was found that network latency is a critical factor that currently can only be reduced at a certain degree.

3.1.4.2 Hybrid Simulation of a Six-Storey Steel Frame with SCED Braces

Civil structures, such as buildings and bridges, are designed to behave in inelastic range due to the design level seismic excitation. The inelasticity of the structure generally lead to residual deformation which greatly impact the usability of the structure. To overcome such issue, there has been active development on self-centering systems which can minimize the residual deformation after an earthquake event. One of such systems is a the self-centering energy-dissipating (SCED) brace which has been developed by Christopoulos et al. (2008) and Erochko et al. (2015). The SCED brace consists of post-tensioned steel tubular sections with energy-dissipating mechanism in the system.

The seismic performance of a six-storey building structure with SCED braces was evaluated with a hybrid simulation (Kammula et al. 2014). The building structure was designed for Los Angeles. Moment resisting frames were used as the lateral load resisting system in the east-west direction while the SCED braces were used in the north-south direction. Figure 3.7 shows the plan and elevation of the building structure. In the hybrid simulation, which was carried out using UI-SimCor (Kwon et al. 2008), one of the braces was modelled in the laboratory while the rest of the structural system was modelled in OpenSees. The network interface program for controllers, NICON, (Zhan and Kwon 2015) was used to interface NI hardware (A/D and D/A converters) with the main integration module, UI-SimCor. As a network interface for OpenSees, the Network Interface for Console Applications

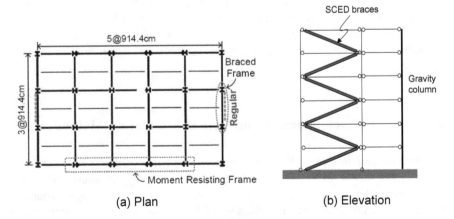

(a) Plan   (b) Elevation

Fig. 3.7 Overall configuration of the building structure

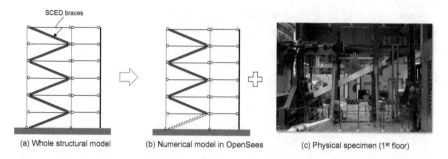

(a) Whole structural model   (b) Numerical model in OpenSees   (c) Physical specimen (1st floor)

Fig. 3.8 Substructuring of the building for hybrid simulation configuration (Kammula et al. 2014)

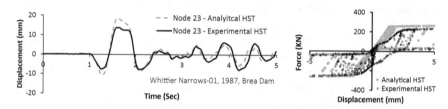

Fig. 3.9 Comparison of numerical prediction and hybrid simulation (Kammula et al. 2014)

(NICA) was used which communicated with OpenSees through named pipes. Figure 3.8 shows how the building is substructured for hybrid simulation.

The results from the hybrid simulation is compared with the numerical prediction in Fig. 3.9. The two results show approximately 35% of difference in the response of the 1st floor. Post-experiment study revealed that the difference mainly results from the energy dissipation around the origin of the hysteretic curve (Fig. 3.9b). In addition, the slight difference in the post-yield stiffness in the compression direction also contributed to the difference. Because the specimen is not damaged due to the inelastic deformation, close to forty hybrid simulations were carried out to develop the fragility of the structure with SCED braces through hybrid simulations. In addition, the impact of the selection of the physically tested element on the hybrid simulation result was investigated, which will be further discussed in Sect. 3.2.

3.2 Latest Development in the Hybrid Simulation Method

A hybrid simulation method is a very attractive simulation method which can take advantage of unique experimental and numerical modelling approaches. Yet, the simulation method still has a large room for improvement for general application in research and engineering practice. Section 3.2.1 discusses some of the limitations and challenges in hybrid simulations. Section 3.2.2 presents a model updating method, which aims to address one of the challenges presented in Sect. 3.2.1. In Sect. 3.2.3, a recently developed hardware, which can test up to ten uniaxially loaded physical specimens in a hybrid simulation, is presented.

3.2.1   Limitations and Challenges

Hybrid simulations can integrate one or more physical specimens with a numerical model. Main objective of such integration is to realistically represent the cyclic hysteretic behaviour of critical structural elements in a structural system such that the global structural response can be more accurately predicted. A hybrid model (i.e. a model that combines physical and numerical substructures) is more realistic than a purely numerical model because it includes at least one real specimen. Yet, the improvement in the accuracy in prediction greatly depends on the contribution of the element to overall response of the structure. For example, if a five-storey structure consists of 20 buckling restrained braces (i.e. 4 braces per floor), modelling only one of them as a physical specimen may not increase the accuracy of the prediction. In addition, there is always an issue of which element should be tested if only a few of many elements in a structural system can be physically modelled.

For example, in the hybrid simulation of the six-storey building with SCED braces in Sect. 3.1.4, six different cases were tested to investigate the impact of the physically tested element on the global response. In each case, the specimen represented the braces at different floor. The first case, where the physical specimen represents the brace on the first floor is assumed to be a reference case. Then, the responses from other cases, where the specimen represents the brace on other floors, are compared with the reference case. The maximum responses of 1st, 3rd, and 5th floors are compared in Fig. 3.10. The figure clearly shows that the selection of physically tested specimen clearly influence the global structural response. It is difficult to tell which case is the most close to the actual response because it depends on the frequency content of input excitation and the inelastic response of structural elements. However, the results clearly show that the hybrid simulation (or physically tested specimen) influences the structural response and the selection of physically tested element requires a better systematic strategy.

On the other hand, if a structural response is dominated by only a few critical elements, then a hybrid simulation is an ideal testing method. For example, if a

Fig. 3.10 Impact of element selection on the maximum drift (Kammula 2012)

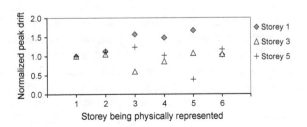

building is supported on a base-isolation system which allows the building to remain in the elastic range, then the global response of the building will be dominantly influenced by the hysteretic behaviour of the base isolator. The response of a high-rise building with tuned liquid damper or tuned mass damper will also dominated by the hysteretic behaviour of dampers, when the building is subjected to wind load. In these scenarios, the hybrid simulation will be an ideal simulation method to predict the response of the structural system.

To overcome some of the limitations of hybrid simulations discussed in the above, several different approaches have been taken. In the first approach, the measured hysteretic behaviour of physically tested elements are used to improve the accuracy of numerical model through a model updating method (Elanwar and Elnashai 2014; Kwon and Kammula 2013; Song and Dyke 2013; Wu et al. 2016). In the second approach, a dedicated experimental system is developed to test many physical elements simultaneously in a hybrid simulation (Mojiri et al. 2015b). These recent developments are briefly introduced in Sect. 3.2.2 and Sect. 3.2.3, respectively.

3.2.2 Model Updating Method

Model updating methods for hybrid simulation have been actively developed to improve the accuracy of hybrid simulations when a limited number of elements are experimentally represented. Several different approaches have been proposed. (Elanwar and Elnashai 2014), for example, updated parameters of material's constitutive model based on the experimentally observed behaviour. Wu et al. (2016) proposed a method where the idealized sectional behaviour is updated. The response of nonlinear constitutive models are path-dependent. Thus, changing model parameters during a simulation may lead to unexpected model behaviour. Thus, rather than updating model parameters, (Kwon and Kammula 2013) proposed a method in which the restoring force from a numerical model is defined based on a weighted average of several alternative numerical models. Song and Dyke (2013) proposed a model-updating method which is applicable to real-time hybrid simulation. In the following, conceptual background of the method proposed by Kwon and Kammula (2013) is presented.

The Eq. (3.1) in Sect. 3.1.1 is numerically solved in a hybrid simulation. In a substructure hybrid simulation, the restoring force term, $\mathbf{r}$, consists of a contribution of a physical specimen to the restoring force, $\mathbf{r}_E$, and a contribution of a numerical model to the restoring force, $\mathbf{r}_N$ as expressed in Eq. (3.2). To evaluate $\mathbf{r}_N$ in a conventional numerical analysis, the most feasible numerical model is selected. Yet, the selection of a numerical model and its modeling parameters may not capture the actual behaviour of a structure unless the model is calibrated or validated against experimental results. In many situations, especially when newly developed structural elements are tested, development of a numerical model involves a large number of simplifications, assumptions, and engineering judgement.

Considering the uncertainties in the accuracy of a numerical model, (Kwon and Kammula 2013) proposed that several alternative numerical models may be used to represent a structural system. The alternative models may cover a wide range of modelling parameters or even different types of numerical elements (e.g. lumped plasticity vs. distributed plasticity model). Then, the contribution of numerical models toward the force vector can be determined based on weighted average of the restoring forces from the multiple alternative models, i.e.

$$\mathbf{r}_N = \sum_{i=1}^{m} w_i \mathbf{r}_{N,i} \tag{3.18}$$

where m is the number of alternative numerical models, i is the index for the alternative numerical models, $\mathbf{r}_{N,i}$ is the restoring force vector from an alternative numerical model i, and w_i is a weighting factor for model i. Then, the Eq. (3.2) can be rewritten as

$$\mathbf{r} = \mathbf{r}_E + \sum_{i=1}^{n} w_i \mathbf{r}_{N,i} \tag{3.19}$$

In an extreme case, if one of the alternative numerical models perfectly replicates the experimental behaviour, the weighting factor of the model is 1 while all other models' weighting factor is zero.

To use Eq. (3.18) as a restoring force from a numerical model, it is necessary to define the weighting factors at each time step. Kwon and Kammula (2013) proposed that a similar set of numerical models for the physically tested specimen can be used to define the weighting factors. Figure 3.11 presents illustration of the proposed method. In this figure, the pairs of numerical models (N_1, $E_{N,1}$), (N_2, $E_{N,2}$), etc. share identical modelling parameters. The weighting factors of EN,1, EN,2, ..., EN,m, can be optimized based on the experimental substructure, E. Then, the weighting factors can be applied in Eq. (3.19). The method was numerically

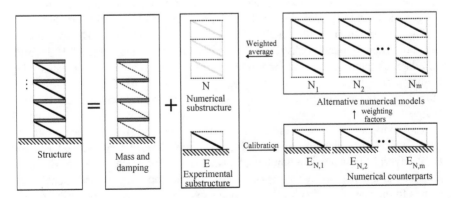

Fig. 3.11 Conceptual illustration of the model updating method (Kwon and Kammula 2013)

verified using Bouc-Wen-Baber-Noori model (Baber and Noori 1985), then experimentally validated.

The method still needs improvement in the optimization method for the weighting factors. Depending on the types of nonlinear hysteretic model, it might be more important to capture the measured force history at large amplitude response (i.e. when the structure develops yielding) than at small amplitude response. In addition, this approach requires several alternative numerical models of the entire system, and communication between the models to impose target displacements and restoring forces. Thus, implementation and application of this method requires some time and efforts, and computing time.

3.2.3 Multi-specimen Hybrid Simulation

One of the methods to overcome the limitations of the hybrid simulation discussed in Sect. 3.2.1 is testing many structural elements in hybrid simulation. For this purpose, (Mojiri et al. 2015b) has developed an experimental system, UT10 Hybrid Simulator, which can test up to ten uni-axially loaded structural elements such as braces, friction dampers, etc. In a typical testing facility, dedicating ten actuators for such testing need is very difficult because not many testing facility is equipped with ten identical actuators. Moreover, the actuators are general testing equipment which need to be used to support various other projects.

The development of the UT10 Hybrid Simulator was possible because it is built upon an existing unique testing equipment, referred to as Shell Element Tester (SET) which was designed to test reinforced concrete shell element (Kirschner and Collins 1986). The SET consists of total 60 hydraulic actuators with 800 kN of force capacity. In the original design, the SET was operated based on oil pressure. In mid-2000s, the equipment was upgraded with servo-controlled hydraulic actuators such that either the stroke or force can be controlled based on PID control. Because this equipment is already equipped with sufficient number of hydraulic actuators, (Mojiri et al. 2015a) developed a loading frame and control program to convert the SET to the ten-element hybrid simulator (UT10 Hybrid Simulator).

The configuration of the loading frame is illustrated in Fig. 3.12 alongside a photo of actual fabrication. As it can be observed from the figure, the UT10 Hybrid Simulator uses ten actuators at the top to control the deformation of specimens. The loading frame is required to impose the loading only in axial direction. Several other actuators are used to provide rigid supports to the specimen and the loading frame.

To provide an interface to the controller of the actuators, the interface program, NICON, was customized for this experimental setup (Block 3.1 in Fig. 3.2). For the UT10 Hybrid Simulator, the number of channels was increased up to ten. External loop was implemented to compensate the error in imposing displacement commands to the specimen. With the error compensation loop, the deformation of the specimen could be controlled below 0.05 mm of error which is sufficiently accurate

Fig. 3.12 UT10 Hybrid Simulator for multi-specimen hybrid simulation (Mojiri et al. 2015a, b)

for the planned hybrid simulations. The flow rate of the hydraulic oil is slow for this equipment. Thus, the equipment supports only pseudo-dynamic hybrid simulations.

Several hybrid simulations were carried out using a hollow steel section, which was used to test the performance of the system. In addition, an adjustable buckling restrained brace was also designed and tested. The simulation results matched well with a numerical prediction. For further references on the development of the UT10 simulator, references are made to (Mojiri et al. 2015a, b).

3.3 Summary

Hybrid simulation has been developed over the past four decades. Several advancement in other fields, such as the widespread of the Internet network since 1990s, low-cost customizable electronic devices (A/D, D/A converters, etc.), and ever-increasing computing speed, a significant development has been made in the hybrid simulation method.

Section 3.1 is intended to provide an overview of the hybrid simulation with some details on widely adopted numerical integration scheme for hybrid simulations, implementation related issues, and application examples. Because the methodology is still being actively developed, and because the specific details on how to implement the method greatly varies from one research group to another, efforts were made to present the methodology as a framework which can encompass widely different implementation methods.

Section 3.2 presented limitation and challenges, and recent progresses in the pseudo-dynamic simulation. As discussed in Sect. 3.2.1, the hybrid simulation method is a very attractive simulation method. Yet, it still requires improvement for application in diverse structural engineering problems. Two recent improvements in pseudo-dynamic hybrid simulations are presented in Sects. 3.2.2 and 3.2.3.

This chapter includes extensive, but not exhaustive, references such that further details can be found from the references. Overview of hybrid simulations were recently published in two recent review papers, (McCrum and Williams 2016; Shao and Griffith 2013), which are good resources to understand the state-of-art on the hybrid simulation methods.

References

Ahmadizadeh M, Mosqueda G, Reinhorn AM (2008) Compensation of actuator delay and dynamics for real-time hybrid structural simulation. Earthq Eng Struct Dyn 37(3.1):21–42. doi:10.1002/eqe.743

Baber TT, Noori MN (1985) Random vibration of degrading, pinching systems. J Eng Mech 111 (3.8):1010. doi:10.1061/(ASCE)0733-9399(1985)111:8(1010)

Bousias S, Kwon O, Evangeliou N, Sextos A (2014) Implementation issues in distributed hybrid simulation. In: 6th world conference on structural control and monitoring, Barcelona, Spain, p 8

Bursi OS, He L, Lamarche C-P, Bonelli A (2010) Linearly implicit time integration methods for real-time dynamic substructure testing. J Eng Mech 136(3.11):1380–1389. doi:10.1061/(ASCE)EM.1943-7889.0000182

Carrion JE, Spencer Jr BF (2007) Model-based strategies for real-time hybrid testing, Urbana. http://hdl.handle.net/2142/3629

Chae Y, Kazemibidokhti K, Ricles JM (2013) Adaptive time series compensator for delay compensation of servo-hydraulic actuator systems for real-time hybrid simulation. Earthq Eng Struct Dyn 42(3.11):1697–1715. doi:10.1002/eqe.2294

Chae Y, Ricles J, Sause R (2014) Large-scale real-time hybrid simulation of a three-story steel frame building with magneto-rheological dampers. Earthq Eng Struct Dyn. doi:10.1002/eqe

Chang S-Y (2010) Explicit pseudodynamic algorithm with improved stability properties. J Eng Mech 136(3.5):599–612. doi:10.1061/(ASCE)EM.1943-7889.99

Chen C (2007) Development and numerical simulation of hybrid effective force testing method. Ph.D. Dissertation, Lehigh University, Ann Arbor.

Chen C, Ricles JM (2008) Development of direct integration algorithms for structural dynamics using discrete control theory. J Eng Mech 134(3.8):676. doi:10.1061/(ASCE)0733-9399(2008) 134:8(676)

Chen C, Ricles JM (2009) Improving the inverse compensation method for real-time hybrid simulation through a dual compensation scheme. Earthq Eng Struct Dyn 38(3.10):1237–1255. doi:10.1002/eqe.904

Christopoulos C, Tremblay R, Kim H-J, Lacerte M (2008) Self-centering energy dissipative bracing system for the seismic resistance of structures: development and validation. J Struct Eng 134(3.1):96. doi:10.1061/(ASCE)0733-9445(2008)134:1(96)

Combescure D, Pegon P (1997) α-Operator splitting time integration technique for pseudodynamic testing error propagation analysis. Soil Dyn Earthq Eng 16(7–8):427–443. doi:10.1016/S0267-7261(97)00017-1

Darby AP, Williams MS, Blakeborough A (2002) Stability and delay compensation for real-time substructure testing. J Eng Mech 128(3.12):1276. doi:10.1061/(ASCE)0733-9399(2002)128:12(1276)

Dermitzakis SN, Mahin SA (1985) Development of substructuring techniques for on-line computer controlled seismic performance testing. Earthquake Engineering Research Center, University of California, Berkeley

Dimig J, Shield C, French C, Bailey F, Clark A (1999) Effective force testing: a method of seismic simulation for structural testing. J Struct Eng 125(3.9):1028–1037. doi:10.1061/(ASCE)0733-9445(1999)125:9(1028)

Elanwar HH, Elnashai AS (2014) On-line model updating in hybrid simulation tests. J Earthq Eng 18(3.3):350–363. doi:10.1080/13632469.2013.873375

Elnashai AS, Spencer BF, Kim SJ, Holub CJ, Kwon OS (2008) Hybrid distributed simulation of a bridge-foundation-soil interacting system. In: The 4th international conference on bridge maintenance, safety, and management, Seoul, Korea

Erochko J, Christopoulos C, Tremblay R (2015) Design and testing of an enhanced-elongation telescoping self-centering energy-dissipative brace. J Struct Eng 141(3.6):04014163. doi:10.1061/(ASCE)ST.1943-541X.0001109

Hakuno M, Shidawara M, Hara T (1969) Dynamic destructive test of a cantilever beam controlled by an analog-computer. Trans Jpn Soc Civil Eng 171(171):1–9

Horiuchi T, Inoue M, Konno T, Namita Y (1999) Real-time hybrid experimental system with actuator delay compensation and its application to a piping system with energy absorber. Earthq Eng Struct Dyn 28(3.10):1121–1141. doi:10.1002/(SICI)1096-9845(199910)28:10<1121::AID-EQE858>3.0.CO;2-O

Huang X, Kwon O-S (2015) Development of integrated framework for distributed multi-platform simulation. In: 6AESE/11ANCRiSST, Champaign, IL

Kammula V (2012) Application of hybrid simulation to fragility assessment of self-centering energy dissipative (SCED) bracing system. University of Toronto

Kammula V, Erochko J, Kwon OS, Christopoulos C (2014) Application of hybrid-simulation to fragility assessment of the telescoping self-centering energy dissipative bracing system. Earthq Eng Struct Dyn. doi:10.1002/eqe.2374

Karavasilis TL, Seo C-Y, Ricles J (2008) HybridFEM: a program for dynamic time history analysis of 2D inelastic framed structures and real-time hybrid simulation (No. ATLSS Report No. 08-09), Bethlehem, PA. http://www.nees.lehigh.edu/wordpress/uploads/reports/HybridFEM-2D_4.2.4_Users_Manual.pdf

Kim SJ, Christenson R, Phillips B, Spencer Jr BF (2012) Geographically distributed real-time hybrid simulation of MR dampers for seismic hazard mitigation. In: 20th analysis and computation specialty conference. American Society of Civil Engineers, Reston, VA, pp 382–393. doi:10.1061/9780784412374.034

Kirschner U, Collins MP (1986) Investigating the behaviour of reinforced concrete shell elements (No. Publication No. 86-9), Toronto

Kwon O-S, Elnashai AS (2008) Seismic Analysis of Meloland road overcrossing using multiplatform simulation software including SSI. J Struct Eng 134(3.4):651–660. doi:10.1061/(ASCE)0733-9445(2008)134:4(651)

Kwon OS, Elnashai AS, Spencer BF (2008) A framework for distributed analytical and hybrid simulations. Struct Eng Mech 30(3.3):331–350. doi:10.12989/sem.2008.30.3.331

Kwon OS, Kammula V (2013) Model updating method for substructure pseudo-dynamic hybrid simulation. Earthq Eng Struct Dyn 42(3.13):1971–1984. doi:10.1002/eqe.2307

Kwon O-S, Nakata N, Elnashai AS, Spencer BF (2005) Technical note: a framework for multi-site distributed simulation and application to complex structural systems. J Earthq Eng 9(3.5):741–753. doi:10.1080/13632460509350564

Magonette G (2001) Development and application of large-scale continuous pseudo-dynamic testing techniques. Philos Trans Royal Soc A: Math Phys Eng Sci 359(1786):1771–1799. doi:10.1098/rsta.2001.0873

Mahin SA, Shing PB (1985) Pseudodynamic method for seismic testing. J Struct Eng 111 (3.7):1482–1503. doi:10.1061/(ASCE)0733-9445(1985)111:7(1482)

Mahmoud HN, Elnashai AS, Spencer BF, Kwon O, Bennier DJ (2013) Hybrid simulation for earthquake response of semirigid partial-strength steel frames. J Struct Eng 139(3.7):1134–1148. doi:10.1061/(ASCE)ST.1943-541X.0000721

Maynard K, Kwon O-S (2014) Network connection quality of service assessment for real-time hybrid simulation applications, Toronto

Mazzoni S, McKenna F, Scott MH, Fenves GL (2007) The OpenSees command language manual, version 2.0. Pacific Earthquake Engineering Research Center, University of California at Berkeley, Berkeley, CA. http://opensees.berkeley.edu

McCrum DP, Williams MS (2016) An overview of seismic hybrid testing of engineering structures. Eng Struct 118:240–261. doi:10.1016/j.engstruct.2016.03.039

Mojiri S, Huang X, Kwon O-S, Christopoulos C (2015a) Design and development of ten-element hybrid simulator and generalized substructure element for coupled problems. In: VI international conference on computational methods for coupled problems in science and engineering, Venice, Italy, p 12

Mojiri S, Kwon O-S, Christopoulos C (2015b) Development of 10-element hybrid simulator and its application to seismic performance assessment of structures with hysteretic energy dissipative braces. In: 6AESE/11ANCRiSST, Champaign, IL

Mosqueda G, Ahmadizadeh M (2011) Iterative implicit integration procedure for hybrid simulation of large nonlinear structures. Earthq Eng Struct Dyn 40(3.9):945–960. doi:10.1002/eqe.1066

Mosqueda G, Stojadinovic B, Mahin SA (2004) Geographically distributed continuous hybrid simulation. In: 13th world conference on earthquake engineering, Vancouver, BC, Canada

Mosqueda G, Stojadinovic B, Hanley J, Sivaselvan M, Reinhorn A (2008) Hybrid seismic response simulation on a geographically distributed bridge model. J Struct Eng ASCE 134 (3.4):535–543

Mosqueda G, Stojadinovic B, Hanley J, Sivaselvan M, Reinhorn AM (2008) Hybrid seismic response simulation on a geographically distributed bridge model. J Struct Eng 134(3.4):535–543. doi:10.1061/(ASCE)0733-9445(2008)134:4(535)

Murray JA, Sasani M, Shao X (2015) Hybrid simulation for system-level structural response. Eng Struct 103:228–238. doi:10.1016/j.engstruct.2015.09.018

Nakashima M, Akazawa T, Sakaguchi O (1993) Integration method capable of controlling experimental error growth in substructure pseudo dynamic test. AIJ: J Struct Constr Eng 454 (61–71)

Nakashima M, Masaoka N (1999) Real- time on-line test for MDOF systems. Earthq Eng Struct Dyn 28:393–420

Negro P, Pinto AV, Verzeletti G, Magonette GE (1996) PsD test on four-story R/C building designed according to eurocodes. J Struct Eng 122(3.12):1409–1471

Newmark NM (1959) A method of computation for structural dynamics. J Eng Mech ASCE 85 (EM3):67–94

Ojaghi M, Williams MS, Dietz MS, Blakeborough A, Lamata Martínez I (2014) Real-time distributed hybrid testing: coupling geographically distributed scientific equipment across the Internet to extend seismic testing capabilities. Earthq Eng Struct Dyn 43(3.7):1023–1043. doi:10.1002/eqe.2385

Pan P, Tada M, Nakashima M (2005) Online hybrid test by internet linkage of distributed test-analysis domains. Earthq Eng Struct Dyn 34(3.11):1407–1425. doi:10.1002/eqe.494

Pan P, Tomofuji H, Wang T, Nakashima M, Ohsaki M, Mosalam KM (2006) Development of peer-to-peer (P2P) internet online hybrid test system. Earthq Eng Struct Dyn 35(3.7):867–890. doi:10.1002/eqe.561

Sadeghian V, Vecchio F, Kwon O (2015) An integrated framework for analysis of mixed-type reinforced concrete structures. In: CompDyn 2015, Crete, Greece, p 13

Saouma V, Kang D-H, Haussmann G (2012) A computational finite-element program for hybrid simulation. Earthq Eng Struct Dyn 41(3.3):375–389. doi:10.1002/eqe.1134

Schellenberg A, Huang Y, Mahin SA (2008) Structural FE-software coupling through the experimental software framework, OpenFresco. In: The 14th world conference on earthquake engineering, Beijing, China

Schellenberg A, Mahin S, Fenves G (2009) Advanced implementation of hybrid simulation

Sextos AG, Bousias S, Taskari O, Evangeliou N, Kwon O, Elnashai A et al (2014) An intercontinental hybrid simulation experiment for the purposes of seismic assessment of a three-span R/C bridge. In: 10th national conference on earthquake engineering, Anchorage, Alaska

Shao X, Griffith C (2013) An overview of hybrid simulation implementations in NEES projects. Eng Struct 56:1439–1451. doi:10.1016/j.engstruct.2013.07.008

Shing P-SB, Vannan MT, Cater E (1991) Implicit time integration for pseudodynamic tests. Earthq Eng Struct Dyn 20(3.6):551–576. doi:10.1002/eqe.4290200605

Simulia D (2014) ABAQUS Version 6.14, analysis user's guide

Song W, Dyke S (2013) Real-time dynamic model updating of a hysteretic structural system. J Struct Eng 130424204433003. doi:10.1061/(ASCE)ST.1943-541X.0000857

Spencer Jr BF, Elnashai A, Kuchma D, Kim S, Holub C, Nakata N (2006) Multi-site soil-structure-foundation interaction test (MISST). University of Illinois at Urbana-Champaign

Spencer B, Finholt T, Foster I, Kesselman C, Beldica C, Futrelle J et al (2004) NEESGRID : a distributed collaboratory for advanced earthquake engineering experiment and simulation. In: 13 th world conference on earthquake engineering, Vancouver, BC, Canada

Wand T, Jacobsen G, Cortes-Delgado M, Mosqueda G (2010) Distributed online hybrid test of a four-story steel moment frame using flexible test scheme. In: 9th US conference on earthquake engineering, Toronto, ON

Takahashi Y, Fenves G (2006) Software framework for distributed experimental–computational simulation of structural systems. Earthq Eng Struct Dyn 35(3.3):267–291. doi:10.1002/eqe.518

Takanashi K, Ohi K (1983) Earthquake response analysis of steel structures by rapid computer-actuator on-line system, (3.1) a progress report, trial system and dynamic response of steel beams. Inst Bull Earthq Resist Struct Res Center (ERS) I:103–109

Takanashi K, Udagawa K, Seki M, Okada T, Tanaka H (1975) Nonlinear earthquake response analysis of structures by a computer actuator online system. Trans Arch Inst Jpn 229:77–83

Vecchio FJ, Wong P (2003) VecTor2 and FormWorks manual

Watanabe E, Kitada T, Kunitomo S, Nagata K (2001) Parallel pseudodynamic seismic loading test on elevated bridge system through the Internet. Singapore

Whyte C, Wotherspoon L, Kim H, Stojadinovic B, Ma Q (2010) Time step optimization for distributed hybrid simulation between University of California—Berkeley and University of Auckland. In: 9th US Conference on Earthquake Engineering, Toronto, ON

Wu B, Chen Y, Xu G, Mei Z, Pan T, Zeng C (2016) Hybrid simulation of steel frame structures with sectional model updating. Earthq Eng Struct Dyn 45(3.8):1251–1269. doi:10.1002/eqe.2706

Zhan H, Kwon O-S (2015) Actuator controller interface program for pseudo-dynamic hybrid simulation. In: 2015 world congress on advances in structural engineering and mechanics, Songdo, Korea

Zhao J, French C, Shield C, Posbergh T (2006) Comparison of tests of a nonlinear structure using a shake table and the EFT method. J Struct Eng 132(3.9):1473–1481. doi:10.1061/(ASCE)0733-9445(2006)132:9(1473)

Chapter 4
Field Structural Dynamic Tests at the Volvi-Greece European Test Site

George C. Manos

Abstract This paper presents summary results from a series of field experiments conducted in situ with structural models, built with realistic foundation conditions and instrumented in an effort to study their dynamic behaviour at a test site in Greece. Through these experiments the influence of structure-foundation-soil interaction could be studied in some detail. Moreover, the variation of the dynamic characteristics of these structural models could be linked to certain changes in their structural system, including the development of structural damage. The measured response was next utilized to validate numerical tools capable of predicting influences arising from such structural changes as well as from soil-foundation interaction. The studied model structures were supported on soft soil deposits thus allowing the study of structure-foundation-soil interaction effects during low-to-medium intensity man-made excitations. The in situ experiments conducted at the Volvi test site provided measurements that could be used to verify certain hypotheses as well as validated numerical simulations that were developed to predicted the response of the studied structures. A focal point was the investigation of the influence of the masonry infills for the 6-story structure as well as that of the soil-foundation interaction effects for both the 6-story as well as the bridge pier model structures and their inclusion in the numerical simulations. Special study concentrated on how neighbouring structures can introduce coupling in the soil-foundation interaction as well as how soil-foundation interaction effects propagate.

Keywords In situ tests · Dynamic response · System identification · Masonry infills

G.C. Manos (✉)
Department of Civil Engineering, Aristotle University of Thessaloniki,
54124 Thessaloniki, Greece
e-mail: gcmanos@civil.auth.gr

© Springer International Publishing AG 2017
A.G. Sextos and G.D. Manolis (eds.), *Dynamic Response of Infrastructure to Environmentally Induced Loads*, Lecture Notes in Civil Engineering 2,
DOI 10.1007/978-3-319-56136-3_4

4.1 Overview

One of the best ways to study the effects of earthquakes in structures is to study their behaviour in situ. Instrumenting structures in situ one takes advantage of the realistic foundation conditions that cannot be reproduced with such realism at the laboratory. It is not equally easy to take advantage of in situ earthquake ground motions as their occurrence has a considerable degree of uncertainty even for areas of high seismicity. Instead, ambient or man-made excitations are utilized to excite a structure in a desired way and study its response. This paper presents selective results from a long standing effort to measure in situ the dynamic response of model structures of relatively large dimensions at a test site designated for this purpose (Manos et al. 1997, 1998a, b, 2015). The advantage of performing such field

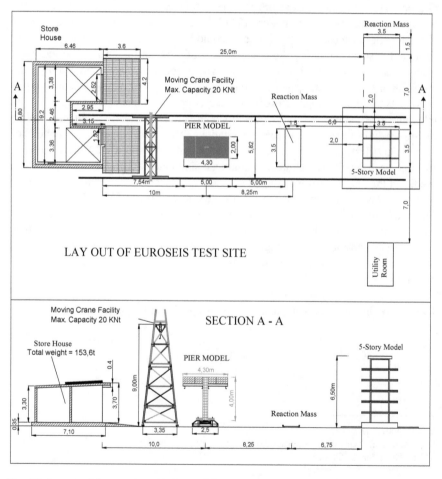

Fig. 4.1 Layout of the test site with the model structures

experiments lies on the fact that the studied structural models are of considerable size and have realistic foundation conditions. Thus, one of the objectives was to study the structure-foundation-soil interaction effects by monitoring the dynamic response of these structural models and their foundations together with the vibrations of the soil in the vicinity of these models. In this way, the obtained measurements could be used to verify assumptions that are utilised in the analysis or design of systems with flexible foundation conditions (Manos et al. 2008; Mylonakis et al. 2002). The model structures were designed so that they could respond to man-made excitations in an easily recorded way as well as develop the desired interaction between the foundation and the surrounding soil. They were also intentionally designed in such a way that certain non-linear mechanisms could develop when man-made excitations reached a feasible medium intensity level or during an earthquake event of medium intensity occurring at a relatively small distance from the test site. The general layout of this test site is depicted in Fig. 4.1. The first structural model (right side of Fig. 4.1), is a 1:3 scaled physical simulation of a 5-story reinforced concrete (R/C) building, built in 1994, with an added 6th story steel apex, built in 1997. The second model structure, built in 2004, is a small scale representation of a single bridge pier together with a part of its deck that carries additional weight and its foundation block, to investigate the dynamic response of such a structural component (Pinto 1996; Kawashima 2000; Manos et al. 2004, 2005, 2015).

4.2   Description of the 6-Story Model Structure

The overall dimensions of this model are 3.5 m × 3.5 m in plan and 6.5 m in height. Its weight together with the foundation slab and the added load, placed at each slab, is approximately 350 KN. This model is similar to a structure located at Chiba Field Station of the I.I.S. of the University of Tokyo in Japan (Okada and Tamura 1992). Figure 4.2 depicts the basic details of this model that is made of reinforced concrete cast in situ.

Only basic information is given here as further details can be obtained from the literature (Manos 1997, 1998a, b, 2000). The following are the basic structural configurations of this model:

a. R/C structure, No added weight, No masonry infills ("Virgin" structure, September–November 1994).
b. R/C structure with 50 KN added weight but without any masonry infills ("Bare" structure, November 1994–June, 1995). Total dead load at this stage 350KN
c. R/C structure with 50 KN added weight (10 KN/slab) and with masonry infills in all but the ground floor (Masonry scheme 1, July 1995–December 1996).

Fig. 4.2 The 6-story model structure (Masonry scheme 2b, September 1997–January 2004)

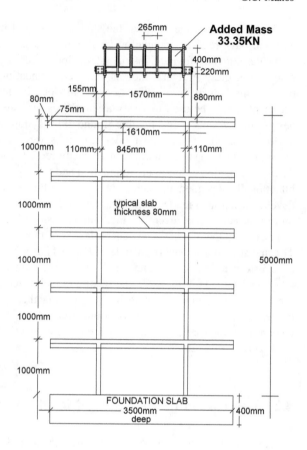

d. R/C structure with 50 KN added weight and with masonry infills in all floors (Masonry scheme 2a, January 1997–August 1997). Total Load at this stage 390 KN.
e. Apart from the existing added weight of 50 KN, extra weight of approximately 33.4 KN is placed on top of a sixth floor specially constructed for this purpose (Masonry scheme 2b, September 1997–January 2004, Fig. 4.3). At this stage the total weight was equal to 423.5 KN.

The above basic configurations were combined with the selected presence of a number of diagonal steel cables at the bays of the story frames to thus form various sub-formations (Table 4.1). For all the listed structural configurations low-level free vibration tests were performed and the corresponding dynamic response was measured.

Fig. 4.3 The 6-story model structure (Masonry scheme 2b, September 1997–January 2004)

Table 4.1 Summary of measured and predicted eigen-frequencies for the Volvi test structure with permanent instrumentation deployed in situ

Description of the model story structural configurations*	1st translational x-x (Hz)	1st translational y-y (Hz)	1st torsional φ-φ (Hz)
5-story-masonry, September 1997 with added mass and masonry infills in all stories (with diagonals)	6.05 {5.666} * x-φ	5.66 {5.570} * y-φ	{6.355}*
6-story-masonry, November 2003 with 6th story extension, no diagonals masonry in all 5 lower stories	4.98 {4.745} *	4.83 {4.673} *	4.91 {5.121}*
6-story-masonry, November 2003 with 6th story extension, with diagonals masonry in all 5 lower stories	5.13 {4.845} *	4.93 {4.745} *	5.13 {5.15}*

{ }* Numerical simulation results

4.3 Low Amplitude Free Vibration Dynamic Tests
of the 6-Story Model Structure

Numerous low-intensity free vibration tests were performed. The permanent instrumentation system consisted of three horizontal acceleration sensors mounted at the mid-plane of the each story slab and the foundation block, as shown in Fig. 4.4; additional vertical acceleration sensors were mounted on the foundation block and the 5th story slab. Moreover, portable equipment was utilized during the low-amplitude dynamic tests. All measured response was analyzed by obtaining the fast Fourier transforms and combined in all measuring points in order to extract the mode shapes and eigen-frequencies. Summary results of this data analysis from a large number of in situ measurements are given by Manos et al. (1998b, 2000). These results are also listed in Table 4.1 together with the corresponding numerical predictions. The numerical simulations included the influence of the masonry infills and the interaction between the masonry infills and the surrounding concrete frame.

The numerical modeling of the masonry infills utilized measurements and observations of an extensive laboratory investigation that was conducted in parallel using one-bay one-story masonry infilled frames constructed with the same dimensions, structural details and material properties as this model structure in situ (Manos et al. 1998a, 2012a). In addition to numerically modeling the R/C frame masonry infill interaction, the flexibility of the foundation was also numerically simulated utilizing linear springs with properties derived from the soil characteristics in situ. The upper soil layer (4 m deep) had a shear way velocity of 135 m/s,

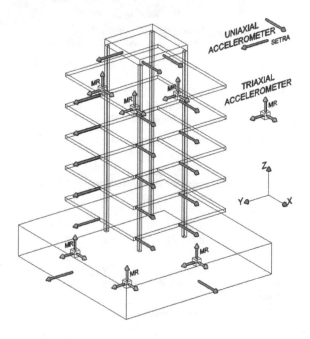

Fig. 4.4 Permanent instrumentation of the 6-story structure

Fig. 4.5 Measured main
translational eigen-modes of
the 6-story model structure
including the soil-foundation
interaction

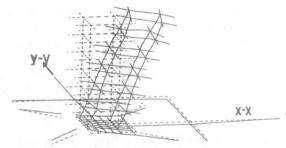

Measured Response (Pull-out x-x 6th Story)
1st Mode **5.005Hz**
mainly translational x-x with rocking at base
6-story Structure with masonry infills in all 5 lower floors (June 1998)

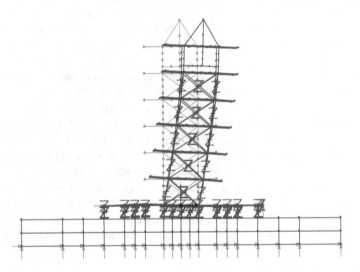

Fig. 4.6 Numerically predicted 1st x-x translational eigen-mode of the 6-story structure. Eigen-frequency = 5.142 Hz

density equal to 2.05 tn/m, and a shear modulus value equal to 37.4 Mpa (Pitilakis et al. 1995, 1999). Figures 4.5 and 4.6 depict the measured and numerically predicted main translational eigen-mode and eigen-frequency with the soil-foundation interaction.

The following summarize the most important observations:

• The presence of the masonry infills influences both the eigen-frequency values and mode shapes, initiates noticeable rocking at the soil-foundation interface and introduces coupled (translational-torsional) modes. When masonry infills are in all but the ground floor (pilotis) the ground floor (being more flexible) exhibits increased deformation demands.

- The numerical predictions are quite successful in capturing the change of stiffness introduced by both the masonry infills as well as the foundation flexibility. It must be pointed out that this numerical modeling made use of laboratory measurement of the masonry stiffness. Moreover, the measured stiffness properties of the soil layers were also made use of.

4.4   Excitation of the 6-Story Structure by Explosions

Apart from the low-intensity free vibration tests this model structure was also excited by explosives being ignited at a distance of approximately 150 m from this structure. The permanent instrumentation captured both the excitation and response of this model structure. The acceleration measurements together with the measured masses were utilized to obtain the overall response in terms of overturning moments and base shear and torque. Additional instrumentation was placed at the soil surface at three locations surrounding the foundation block employing tri-axial accelerometers. The first location for such a sensor was at the y-y axis and at a distance of 5.2 m from the edge of the foundation block (Fig. 4.5).

The second location was at the x-x axis and at a distance of 3.7 m from the edge of the foundation block whereas a third location was at the diagonal between the x-x and the x-x axis and at a distance of 5 m from the edge of the foundation block. Figure 4.7 depicts the measured horizontal acceleration measured at the foundation block as well as at the 5th story slab due to the explosion. The symbol x-x, or y-y denotes that the sensor was located at the x-x axis and measured the horizontal acceleration along the same axis. The explosion initiated a high frequency horizontal acceleration response of the foundation block during the first 1.5 s (Fig. 4.7). The model structure was excited mainly in its translational modes (either x-x or y-y)

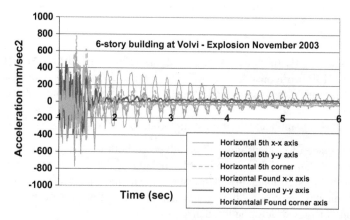

Fig. 4.7 Horizontal acceleration measured at the foundation block and at the 5-story slab

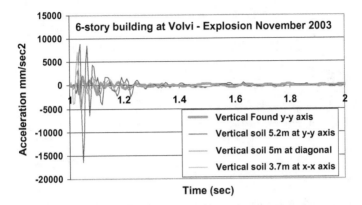

Fig. 4.8 Vertical acceleration measured at the foundation block and at the three soil locations

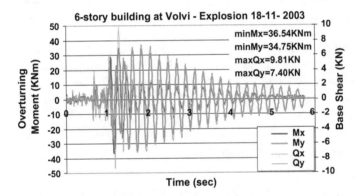

Fig. 4.9 Overturning moment and base shear response of the 6-story structure

and continued to respond in a free-vibration mode for more than 10 s (Fig. 4.7). The horizontal and vertical accelerations of the foundation block were much less than the horizontal and vertical acceleration recorded at the soil surface at a distance approximately 5 m from the foundation block (Fig. 4.8).

The total response, in terms of overturning moment and base shear along the two main axes (x-x and y-y) and the torque, is depicted in Figs. 4.9 and 4.10. It can be seen that the explosion resulted in 36.5 KNm overturning moment and 9.8 KN base shear as maximum absolute values. These values were not sufficient to produce any visible damage to either the masonry infills or the concrete columns or beams. Laboratory testing has shown that more than 40 KN of base shear was required for this purpose (Manos et al. 2012a, b). The soil-foundation interaction represents an additional damping mechanism. From the decay of the free vibration response the equivalent damping ratio of the main translational eigen-modes was estimated to be of the order of 3%. The torsional response of this 6-story model structure was also estimated making use as before of all the horizontal acceleration measurements and

74   G.C. Manos

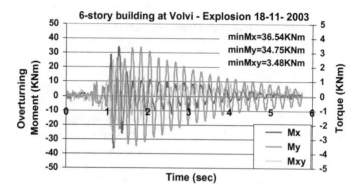

Fig. 4.10 Comparison of the translational with the torsional overall response

the known masses (Fig. 4.10). The explosion excited also the torsional response of this 6-story model and resulted in 3.48KNm maximum absolute torque value; thus, the peak torsional response was one order of magnitude lower than the peak main translational/overturning—rocking response. This torsional response was excited despite the symmetry of the structure and the distributed mass and it must be attributed to the fact that the first torsional eigen-mode had an eigen-frequency value quite close to the main translational eigen-frequency values (see Table 4.1).

4.5 Summary and Conclusions

The in situ experiments at the test site provided measurements that could be used to verify certain hypotheses as well as validated numerical simulations that were developed to predict the response of the studied structures. A focal point was the investigation of the influence of the masonry infills for the 6-story structure as well as that of the soil-foundation interaction effects for both the 6-story as well as the bridge pier model structures and their inclusion in the numerical simulations. Special study concentrated on how neighbouring structures can introduce coupling in the soil-foundation interaction (Renault and Meskouris 2005; Manos et al. 2006, 2007) as well as how soil-foundation interaction effects propagate (Pitilakis and Terzi 2012).

The following summarize the most important observations for the 6-story structure with the masonry infills: (a1) The presence of masonry infills influences significantly both the eigen-frequency values and mode shapes, initiates noticeable rocking at the soil-foundation interface and introduces coupled (translational-torsional) modes. When masonry infills are in all but the ground floor (pilotis) the ground floor (being more flexible) exhibits increased deformation demands (Manos et al. 2000). (b1) The numerical predictions are quite successful in capturing the change of stiffness introduced by both the masonry infills as well as the foundation flexibility (Manos et al. 2000, 2012a, b). This in situ investigation regarding the influence of the masonry infills and its numerical simulation was supplemented with extensive investigation published elsewhere (Manos et al. 2012a, b).

References

Kawashima K (2000) Seismic performance of RC bridge piers in Japan: an evaluation after the 1995 Hyogo-ken nanbu earthquake. Progr Struct Eng Mater 2:82–91

Manos GC et al (1997) The dynamic response of a 5-story R.C. structure in-situ at the European test site at Volvi. 14th SMIRT, Lyon, France, vol H, pp 283–290

Manos GC et al (1998a) Euroseistest Volvi—Thessaloniki, a European test-site for engineering seismology, earthquake engineering and seismology. Final Report to the Commission of the European Communities, Task 2 "Complementary works tests and surveys at EURO-SEISTEST" and Task 3, "Data Analysis and Validation of new and existing codes with Euroseistest data"

Manos GC et al (1998b) The dynamic response of a 5-story structure at the European test site at Volvi-Greece. 6th U.S. National Earthquake Engineering Conference

Manos GC et al (2000) Influence of masonry infills on the earthquake response of multi-story reinforced concrete structures. Presented at the 12th WCEE, Auckland, New Zealand

Manos GC et al (2004) Dynamic and earthquake response of model structures at the Volvi-Greece European test site. 13th WCEE, Vancouver, Canada, No. 787

Manos GC, Renault P, Sextos A (2005) Investigating the design implications of the influence between neighboring model structures at the Euroseistest site. Proceedings of EURODYN 2005

Manos GC, Kourtides V, Soulis V, Sextos A, Renault P (2006) Study of the dynamic response of a bridge pier model structure at the Volvi-Greece European test site. 8th U.S. National Conference on Earthquake Engineering, San Francisco, pp 17–21

Manos GC, Kourtides V, Sextos A, Renault P, Chiras S (2007) Study of the dynamic soil-structure interaction of a bridge pier model based on structure and soil measurements. 9th Canadian Conference on Earthquake Engineering, Ottawa, Ontario, Canada

Manos GC, Kourtides V, Sextos A (2008) Model bridge pier foundation—soil interaction implementing, in-situ/shear stack testing and numerical simulation. 14WCEE, Beijing, China

Manos GC, Soulis VJ, Thauampteh J (2012a) The behavior of masonry assemblages and masonry-infilled R/C frames subjected to combined vertical and cyclic horizontal seismic-type loading. Adv Eng Softw 45:213–231

Manos GC, Soulis VJ, Thauampteh J (2012b) A nonlinear numerical model and its utilization in simulating the in-plane behaviour of multi-story R/C frames with masonry infills. Open Constr Build Technol J 6(Suppl 1-M16):254–277 (2012)

Manos GC, Pitilakis KD, Sextos AG, Kourtides V, Soulis VJ, Thauampteh J (2015) Field experiments for monitoring the dynamic soil-structure-foundation response of model structures at a test site. Journal of Structural Engineering, American Society of Civil Engineers, Special Issue "Field Testing of Bridges and Buildings", D4014012, vol 141, no 1

Mylonakis G, Gazetas G, Nikolaou S, Chauncey A (2002) Development of analysis and design procedures for spread footings. Technical Report, MCEER-02-0003

Okada T, Tamura R (1992) Observation of earthquake response of R.C. weak model structures. Bulletin Institute of Industrial Science, University of Tokyo, No. 18

Pinto AV (ed) (1996) Pseudodynamic and shaking table tests on R.C. bridges. ECOEST PREC*8 Report No. 8

Pitilakis K et al (1995) Euroseistest Volvi—Thessaloniki, a European test-site for engineering seismology, earthquake engineering and seismology. Final Report to the Commission of the European Communities, Task 4, Geophysical and Geotechnical Survey, November 1995

Pitilakis K et al (1999) Geotechnical and geophysical description of EURO-SEISTEST, using field, laboratory tests and moderate strong motion records. J Earthq Eng 3(3):381–409

Pitilakis KD, Terzi V (2012) Chapter 7. Experimental and theoretical SFSI studies in a model structure in Euroseistest. In: Sakr MA, Ansal S (eds) Special topics in earthquake geotechnical engineering. Springer, Netherlands, Dordrecht

Renault P, Meskouris K (2005) A coupled BEM/FEM approach for the numerical simulation of bridge structures. EURODYN 2005, pp 1267–1272

Chapter 5
An Intercontinental Hybrid Simulation Experiment for the Purposes of Seismic Assessment of a Three-Span R/C Bridge

Anastasios G. Sextos and Olympia Taskari

Abstract This study presents the challenges encountered in preparing and conducting hybrid experiments between E.U., U.S. and Canada in the framework of an FP7-funded European project focusing on the study of seismic soil-structure interaction effects in bridge structures. The test involved partners located on both sides of the Atlantic; each one assigned a numerical or a physical module of the sub-structured bridge. More precisely, the seismic response of a recently built, 99 m long, three-span, reinforced concrete bridge is assessed, after sub-structuring it into five structural components (modules); four of them being numerically analyzed in computers located in the cities of Thessaloniki (Greece), Patras (Greece), Urbana-Champaign. IL (U.S.) and Toronto (Canada) while an elastomeric bearing was physically tested in Patras (Greece). The results of the hybrid experiment, the challenges met during all stages of the campaign, as well as the feasibility, robustness and repetitiveness of the intercontinental hybrid simulation test are presented and critically discussed.

Keywords Hybrid testing · Multi-platform simulation · Bridges · Earthquake engineering

A.G. Sextos (✉) · O. Taskari
Division of Structural Engineering, Department of Civil Engineering,
Aristotle University, Thessaloniki 54124, Greece
e-mail: asextos@civil.auth.gr

A.G. Sextos
Department of Civil Engineering, University of Bristol, Queens Building, University Walk,
Bristol BS8 1TR, UK

© Springer International Publishing AG 2017
A.G. Sextos and G.D. Manolis (eds.), *Dynamic Response of Infrastructure to Environmentally Induced Loads*, Lecture Notes in Civil Engineering 2,
DOI 10.1007/978-3-319-56136-3_5

5.1 Introduction

Full scale testing is a realistic way to evaluate the behavior of structures under earthquake loading, as well as to verify the effectiveness of the design or retrofy methods for new or existing earthquake-resistant structures, respectively. Notwithstanding the increasing capabilities of the structural engineering laboratories, factors related to space limitation, equipment capacity, scaling issues and the high operational and maintenance cost of the facilities themselves, often set limits to the problems that can be studied through physical experimentation. On the other hand, the advanced analytical and numerical models that are currently available, have their own limitations in capturing the actual complex seismic behavior of the structures and the phenomena studied. This is even more pronounced in case of complex structural behavior (such as strong material or geometrical non-linearities), non-conventional loading or boundary conditions, or significant soil-structure interaction phenomena. As a result, the analysis capabilities are inevitably limited to solving a specific set of relatively narrow problems primarily at a component level.

Given the above merits and drawbacks of the (experimental and numerical) seismic performance evaluation methods, a challenging concept has been introduced for multi-site, on-line, computer- controlled integrated testing-analysis of complex systems. Depending on the number of the locations used for the implementation of a hybrid test, the simulation technique can be characterized as local or geographically-distributed. This multi-site, Real Time Hybrid Simulation (RTHS) approach has already been developed in the United States for the assessment of complex interacting systems. It is supported by National Science Foundation, through the Network for Earthquake Engineering Simulation (NEES, http://www.nees.org) scheme (Kwon et al. 2005; Pan et al. 2005; Saouma et al. 2012; Takahashi and Fenves 2006) and it aims to raise the limitations related to the laboratory capacities. Within this framework, there is no need for either using a single experimental facility or for satisfying physical proximity for the multiple sub-components. The dynamic response of full scale specimens that are discretized into sub-structures is properly controlled via purpose-specific coordination software. Two such specialized software platforms exist to date, i.e., the OpenFresco (Haussmann 2007; Wang et al. 2011) and SimCor (Spencer et al. 2006). The components (analytical, experimental or a combination of both) are treated on different networked computers and, can thus be located anywhere in the world. Another major advantage of hybrid simulation is that it removes a large source of uncertainty compared to pure numerical simulations, by replacing structural elements with complex non-linear behavior with physical specimens tested on the laboratory test bed. Apparently, drawbacks also exist and are related to the necessity for in-depth knowledge of specialized experimental and analytical tools as well as for considerable programming effort and computational cost.

The same concept has also been successfully applied (Kwon and Elnashai 2008) for the coordination of purely numerical analysis modules (where no physical testing is performed, in contrast to the hybrid simulation application). This, so called, "multi-platform simulation" permits the appropriate selection and combination of different analysis packages, thus enabling the concurrent use of the most sophisticated constitutive laws, element types and features of each package for each corresponding part of the system (i.e., abutments, superstructure and supporting pile groups for instance in the case of a long bridge), depending on the foreseen inelastic material behavior, level and nature of the seismic forces and the geometry of the particular problem. As for the case of Hybrid Simulation though, the computational cost and level of expertise is relatively high compared to a conventional all-inclusive simulation package, plus, its efficiency is network-dependent.

EXCHANGE-SSI (EXperimental and Computational Hybrid Assessment Network for Ground-motion Excited, Soil-Structure Interaction systems) is an EU-funded, 7th Framework Program research project, within which, a number of earthquake engineering centers in Europe and the U.S. collaborate on the application of distributed, hybrid or multi-platform experimentation for the study of seismic soil-structure interaction effects in bridge structures. Although this geographically-distributed analysis and hybrid experimentation is currently well established in the U.S. as well as within few specialized centers in Europe (primarily U.K.-NEES, JRC) and in Taiwan, the challenge here was twofold: (a) to successfully run intercontinental hybrid experiments between European and North American institutions in a fully repetitive manner by overcoming the barrier of the network connection latency introduced while transmitting data through the Atlantic, (b) to physically test a rate-dependent bridge component (i.e., bearing), as opposed to the most common case of a reinforced concrete or a steel structural member and (c) to study a comprehensive, recently constructed, bridge structure. To the best of the author's knowledge, the only similar, successful intercontinental hybrid simulation effort is the test between UC Berkeley and University of Kassel in Germany (Gunnay and Schellenberg 2007), involving a single component.

As the communication time step is the most critical parameter for a successful hybrid experiment, the campaign was set up carefully, initiating from a series of preparatory (numerical-only), intercontinental, multi-platform, multi-partner analyses, as well as, hybrid experiments localized at the University of Patras (Taskari and Sextos 2013). All partners were gradually involved in different stages and in different roles until the performance was optimized in terms of communication time. Ultimately, the bridge was divided into five structural components (modules), each one being analyzed using specific software in a different computer stations (Fig. 5.1) located at Aristotle University of Thessaloniki (AUTH), University of Patras (UPAT), University of Naples/Sannio (USAN), University of Illinois at Urbana-Champaign (UIUC) and University of Toronto (UoT). At the final stage, the numerical module representing the left bridge bearing was replaced by the specimen and was physically tested at the University of Patras. In both cases (i.e., multi-platform and hybrid experimentation) the analysis coordinator SimCor (Kwon et al. 2007), developed at the University of Illinois at Urbana-Champaign was used. The description of the series of the experiments conducted, from the geographically-distributed multi-platform simulation to the intercontinental hybrid

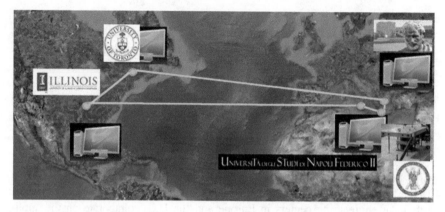

Fig. 5.1 Geographical distribution of the numerical and experimental sub-structures involved in the intercontinental multi-platform and hybrid experiments

experiment, as well as the limitations, challenges met and future developments are discussed in the following.

5.2 Description of the Bridge

A typical three-span overpass (T7) of a total length of 99.0 m which is part of EGNATIA highway in Northern Greece was adopted for study. The two outer spans of the bridge have a length of 27.0 m each, while the middle span is 45.0 m long. The slope of the structure along the bridge axis is constant and equal to 7% with increasing altitudes towards the west abutment. The deck consists of a 10 m wide, prestressed concrete box girder section, while the two piers are designed with a solid circular reinforced concrete section with diameter equal to 2.0 m and are monolithically connected to the deck. The heights of the left and the right pier are 7.95 and 9.35 m, respectively, while two series of 48 longitudinal bars of 25 mm diameter are spaced equally around the section perimeter. The transverse reinforcement consists of an outer spiral of 14 mm diameter spaced at 75 mm and an inner 16 mm spiral equally spaced. The deck is supported on seat type abutments of a backwall height equal to 2.0 m, through two pot bearings (350 mm × 450 mm × 136 mm) that permit sliding along the two principal bridge axes. Sliding joints of 10 cm and 15 cm length separate the deck from the abutment along the longitudinal and the transverse direction, respectively. The foundation rests on surface footings given the stiff soil formations corresponding to class B according to EC8-Part 2 or C according to NEHRP (FEMA440 2004). In particular, the pier footings are 9.0 m long by 8.0 m wide and 2.0 m thick, while the footings supporting the abutments are 12.0 m × 4.5 m × 1.5 m. A general overview of the bridge configuration is illustrated in Fig. 5.2. The bridge was designed to a peak

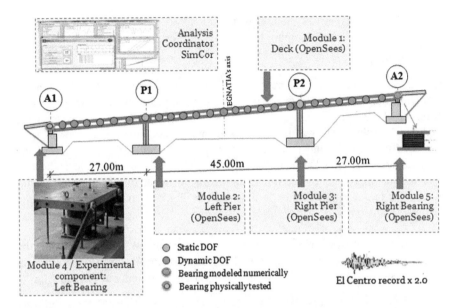

Fig. 5.2 Layout of the bridge substructuring for the multi-platform and the hybrid simulation

ground acceleration of 0.16 g adopting an importance factor equal to 1.0, and a behavior (or force reduction) factor equal to 2.40 according to Greek Seismic Code (EAK2000, E39/99) that was used at the time of construction (Ministry of Public Works of Greece 1999, 2000).

5.3 Computational and Experimental Scheme

The specialized software platform SimCor (Spencer et al. 2006) developed by the research group of the University of Illinois was used for coordinating the preliminary multi-platform analysis used to optimize the envisaged hybrid experiment. SimCor involves an enhanced Matlab-based script which coordinates software or hardware components supporting the NEESgrid Teleoperation Control Protocol (NTCP), as well as TCP-IP connections outside of the NEES system. Analytical models of some parts of the structure or experimental specimens representing specific parts of the same structure, are all considered as super-elements with many DOFs. Specially developed interface programs permit the interaction with different analysis software such as Zeus-NL (Elnashai et al. 2002), OpenSees (McKenna et al. 2002), FedeasLab (Filippou and Constantinides 2004), and ABAQUS (Hibbit and Sorenson 2006). After the initialization step where the connection between the modules is achieved, the stiffness matrix of the whole structure is evaluated using predefined deformation values. The gravity forces are considered during the static loading stage where displacements due to gravity forces are imposed. Finally, SimCor performs Newmark numerical integration as it steps through the seismic record by utilizing the OS method with a modified α-parameter (a-OS method) which applies numerical damping to the undesired oscillations.

System Sub-structuring

The three-span reinforced concrete bridge was divided into five different components (modules) each one analyzed in a different computer station after appropriate definition of the control points at the joint DOFs of interest. At each analysis step, a predefined displacement was imposed by the analysis coordinator and forces were measured to each specific module to establish the initial stiffness matrix of the sub-structured system. The established matrix was then used in the static and dynamic loading stage to determine the desirable target displacements. A brief description of the five modules is illustrated in Fig. 5.2 and is presented in the following.

- Module 1: Consists of the 99.0 m bridge deck and was analyzed using the OpenSees finite element software. The superstructure is expected to remain linear and was thus modeled using linear elastic beam-column elements.

- Module 2: The left pier was numerically analyzed with OpenSees using fiber sections. The stress-strain relationships for the confined and the unconfined concrete were obtained from the Mander et al. model (Mander et al. 1988) while the uniaxial Giuffré-Menegotto-Pinto (Taucer et al. 1991) material with isotropic strain hardening was used for the reinforcement bars. The median design strength of concrete and the yielding strength of reinforcing steel are 35.7 and 550 MPa, respectively.

- Module 3: The right pier of the bridge was also deemed as an individual module, numerically analyzed with OpenSees using fiber sections, as previously.

- Module 4: The bearings at the left end of the deck were considered as an individual module and were first modeled as an elastomeric bearing element whose initial stiffness was calculated based on experimental results of the individually tested bearings. In the subsequent hybrid simulation, the left bearings were physically tested at the University of Patras.

- Module 5: The bearings between the deck and the right abutment were considered as an individual module also numerically analyzed using OpenSees.

UI-SimCor acted as the Analysis Coordinator in all cases.

Experimental Setup

The experimental setup at the Structures Laboratory of the University of Patras consisted of a pair of bearings placed in a back-to-back configuration between two stiff and plates, which were prevented from displacing or rotating (Fig. 5.3). A (nearly) constant vertical load of 240 kN was imposed to the isolators, regardless of the level of applied shear deformation. The 350 mm-in-diameter low damping rubber bearings used (ALGA, Type NB4) consisted of 7 layers of rubber with a

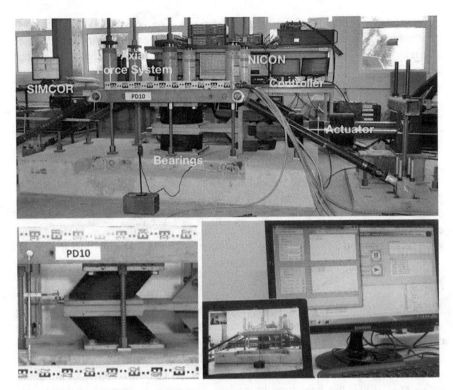

Fig. 5.3 Experimental setup (*top*) and bearings tested (*bottom*, *left*) at the University of Patras along with the computational server at the University of Thessaloniki (*bottom*, *right*)

layer thickness of 11 mm and 6 steel shim plates with a thickness of 6 mm each. The total height of the bearing, including the external connection plates, was 181 mm, while the total rubber height was 77 mm. The prescribed shear modulus of the rubber was 0.99 MPa. The measured horizontal and vertical stiffness of the bearings were estimated as: K_h = 1237 kN/m and K_v = 469.6 MN/m.

5.3.1   Preparatory Computational and Experimental Works

Multi-platform Analysis

Before proceeding with the hybrid experiment, it was deemed necessary to ensure that the multi-platform analysis yields similar results to that of the full model (i.e., the single module finite element model running on a single computer). For that purpose, the bridge was also modeled as a whole in OpenSees. Although it is not presented herein due to space limitations, an excellent match was observed between the sub-structured and the integrated finite element models independently of the geographical distribution of the multiple modules. The optimal geographical distribution and role assignment was identified through successive parametric analyses of a sample four-span, seismically isolated, reinforced concrete bridge (Taskari and Sextos 2013) until the network latency was minimized and the analysis efficiency was improved. From the extensive parametric analyses scheme undertaken, it was seen that among the various sources of analysis delay (i.e., the geographical distribution of modules, the partners' role in the sub-structured analysis, the daytime the simulation took place, as well as pure network connection time), the latter clearly was the most dominating factor. As a result and given the rate-dependency of the bearing that was planned to be physically tested at the University of Patras careful tuning of the involved parameters was made. Finally, the optimum geographical distribution of the modules, as well as the order in which the analysis coordinator was contacting the intercontinental partner modules was identified. The execution of the experiment was also performed within the most efficient time window (12:00 and 14:00 pm Greek time) found to correlate to the lowest network latency between Europe and North America.

Hybrid Simulation

Although a dynamic actuator with a 1500 l/min servovalve was employed for applying the command displacement increments, during the tests presented here the command displacement were applied in a slow, step-wise manner. Owing to the static nature of testing, strain rate effects affecting the bearing response cannot be accounted for. Nevertheless, a force correction procedure (Molina et al. 2002; Palios et al. 2007) has been used in the tests performed to approximately account for this: a "corrected" force is determined as a function of the measured quantities (force, displacement, force rate and displacement rate). The exact dependence was obtained through a series of tests on similar pair of isolators (to avoid any scragging effect) at different testing velocities and for deformation levels similar to those expected during the hybrid tests.

Another issue that had to be dealt with, regards the way in which displacement commands generated by the test coordinator software are introduced as reference signals to the laboratory control system. As the majority of the controllers, even old ones, can accept external input in analog form, the approach implemented at the consortium-partner University of Toronto was used: target displacements sent out by UI-SimCor were received by purpose-built software (Network Interface for Controllers—NICON) under the Labview environment. The software directs the target value to a DAC unit and the (scaled) analog output signal is hard-wired from this unit to the controller as reference displacement or force (analog) value. Upon execution of the command signal, the opposite route was followed. The system was realized at Structures Laboratory and appropriately connected to the laboratory digital controller. Finally, no compensation due to the network (varying) time was introduced in the experimental module as all rate-depending effects on the force response of the isolator were compensated via the characterization process (Molina et al. 2002; Palios et al. 2007).

Table 5.1 Geographical distribution of the modules for all the simulations

	IMPS	HSUPAT	HSGR	IHS
Module 1	AUTH	UPATRAS	AUTH	AUTH
Module 2	UIUC	UPATRAS	AUTH	AUTH
Module 3	AUTH	UPATRAS	AUTH	AUTH
Module 4	UPATRAS	UPATRAS	UPATRAS	UPATRAS
Module 5	U of T	UPATRAS	AUTH	U of T
Coordinator	AUTH	UPATRAS	AUTH	AUTH

Fig. 5.4 Comparison of the force–displacement loops for the three experiments

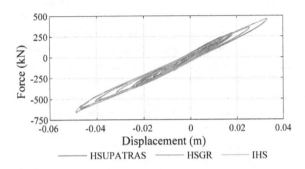

5.4 Series of Multi-platform and Hybrid Experiments

After deciding the geographical distribution of the modules and the experimental setup of the bearings, four types of experiments were conducted among the partners, as summarized in Table 5.1: (a) Intercontinental multi-platform simulation (IMPS), (b) Hybrid simulation at University of Patras (HSUPAT), (c) Hybrid simulation between University of Patras and Aristotle University (HSGR), (d) Intercontinental hybrid simulation (IHS). The ElCentro earthquake record was used for all the aforementioned experiments, with a scale factor of 1.0 and 2.0. A total number of 1000 steps was performed while the time step was set equal to 0.01 s.

5.5 Comparative Assessment of the Dynamic Response Results

Indicative results of the three hybrid experiments (HSUPAT, HSGR, IHS) are presented herein. Figure 5.4 depicts the force-displacement loops for the experimental tested module (left bearing) for the three experiments. It is observed that, despite the system sub-structuring to modules widespread all over the world, the results of the local hybrid experiment (HSUPAT), the Thessaloniki-Patras experiment (HSGR) and the intercontinental one (HIS) are almost identical.

It is noted herein that given the rate-dependency of the bearing that is physically tested at the University of Patras during the hybrid experimentation, the time step delay is an issue of major importance. The time step delay consists of (a) the time required for the finite element analysis at this given step or the progress of the experiment at the same time step, (b) the time required for the analysis coordinator to connect to the various modules and control forces and displacements as well as (c) the pure networking time needed to reach the remote modules worldwide. The disaggregation of the time step delay in the individual modules for the three experiments (IMPS, HSGR, IHS) is presented in Fig. 5.5. It is seen that in the case of the intercontinental multi-platform simulation (IMPS), the numerical part at the most distant module from the analysis coordinator (i.e., Univ. of Illinois at Ubrana-Champaign) in Module 2 is the one that dominates the time required for every time step. As it was expected, in case of the hybrid simulations between both the Greek partners (HSGR) and the entire group of international partners (IHS), the time step delay was primarily due to the experimentally tested component (left bearing). Network delays were kept reasonably low and the tests run successfully with only few time steps exceeding 5 s.

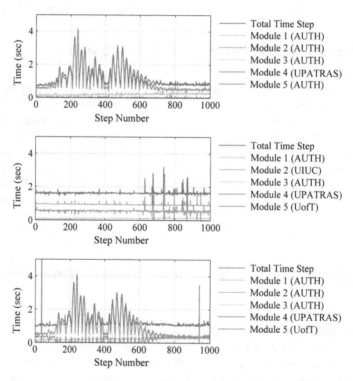

Fig. 5.5 Time step delay for the intercontinental multi-platform simulation (IMPS, *top left*) hybrid simulation between Greek partners (HSGR, *top right*) and the intercontinental hybrid experiment (IHS, *bottom*)

5.6   Conclusions

This study presents the objectives, challenges and performance of an intercontinental hybrid experiment conducted between European and North American partners for the study of the seismic response of a seismically isolated, reinforced concrete bridge structure. The component that was physically tested was the left abutment bearing while the complementary superstructure components were analyzed numerically. The bridge was assessed under moderate and strong earthquake loading, through both geographically-distributed, multi-platform analysis and hybrid simulations, in various combinations until the optimal role allocation was identified. Despite the rate-dependency of the bearing specimen tested and the increased network latency in linking the two sides of the Atlantic, the intercontinental hybrid experiment was accomplished and repeated successfully, highlighting the robustness, efficiency and repetitiveness of the approach.

Acknowledgements This work carried out was funded by the 7th Framework Programme of the European Commission, under the PIRSES-GA-2009-247567-EXCHANGE-SSI grant (Experimental and Computational Hybrid Assessment Network for Ground-Motion Excited Soil-Structure Interaction Systems, http://www.exchange-ssi.net).

References

Elnashai A, Papanikolaou V, Lee D (2002) Zeus NL—a system for inelastic analysis of structures. University of Illinois at Urbana, Champaign
FEMA440 (2004) NEHRP recommended provisions for seismic regulations for new buildings and other structures
Filippou F, Constantinides M (2004) FEDEASLab getting started guide and simulation examples
Gunnay S, Schellenberg A (2007) Continuos intercontinental hybrid testing
Haussmann G (2007) Evaluation of OpenFresco and SIMCOR for fast hybrid single site simulation. In: CU-NEES-07-02, University of Colorado
Hibbit K, Sorenson N (2006) ABAQUS ver. 6.6. User's manual. Pawtucket, USA
Kwon O-S, Elnashai A (2008) Seismic analysis of Meloland road overcrossing using multiplatform simulation software including SSI. J Struct Eng 134(4):651–660
Kwon O-S, Elnashai A, Spencer B, Park K (2007) UI-SIMCOR: a global platform for hybrid distributed simulation. In: 9th Canadian conference on earthquake engineering
Kwon O-S, Nakata N, Elnashai A, Spencer B (2005) A framework for multi-site distributed simulation. J Earthq Eng 9(5):741–753
Mander J, Priestley M, Park R (1988) Theoretical stress-strain model for confined concrete. J Struct Eng 114(8):1804–1826
McKenna F, Fenves G, Scott M (2002) Open system for earthquake engineering simulation. University of California, Berkeley
Ministry of Public Works of Greece (1999) Circular E39/99, Guidelines for the seismic analysis of bridges (in Greek)
Ministry of Public Works of Greece, Greek Seismic Code-EAK 2000, Athens, 2000 (amended June 2003) (in Greek)

Molina FJ, Verzeletti G, Magonette G, Buchet P, Renda V, Geradin M, Parducci A, Mezzi M, Pacchiarotti A, Frederici L, Mascelloni S (2002) Pseudodynamic tests on rubber base isolators with numerical substructuring of the superstructure and the strain-rate effect compensation. Earthq Eng Struct Dyn 31:1563–1582

Palios X, Molina J, Bousias S, Strepelias H, Fardis M (2007) Sub-structured pseudodynamic testing of rate-dependent bridge isolation devices. In: 2nd international conference in advances in experimental structural engineering, Tongji University, Shanghai, China

Pan P, Tada M, Nakashima M (2005) Online hybrid test by internet linkage of distributed test-analysis domains. Earthq Eng Struct Dyn 34(11):1407–1425

Saouma V, Kang D-H, Haussmann G (2012) A computational finite-element program for hybrid simulation. Earthq Eng Struct Dyn 41:375–389

Spencer B, Elnashai A, Park K, Kwon O-S (2006) Hybrid test using UISimCor, Three-site experiment

Takahashi Y, Fenves G (2006) Software framework for distributed experimental–computational simulation of structural systems. Earthq Eng Struct Dyn 35(3):267–291

Taskari O, Sextos A (2013) Robustness, repeatability and resilience of intercontinental distributed computing for the purposes of seismic assessment of bridges. In: 4th international conference on computational methods in structural dynamics and earthquake engineering, Kos, Greece

Taucer F, Spacone E, Filippou F (1991) A fiber beam-column element for seismic response analysis of reinforced concrete structures

Wang Q, Wang J-T, Jin F, Chi F-D, Zhang C-H (2011) Real-time dynamic hybrid testing for soil–structure interaction analysis. Soil Dyn Earthq Eng 31(12):1690–1702

Chapter 6
Experimental Seismic Assessment of the Effectiveness of Isolation Techniques for the Seismic Protection of Existing RC Bridges

L. Di Sarno and F. Paolacci

Abstract The seismic vulnerability assessment of existing and new lifeline systems, especially transportation systems, is becoming of paramount importance in resilient social communities. Unfortunately, the transportation systems, especially in Italy, were mainly built in the late 60s and early 70s and were designed primarily for gravity loads. As a result, most of the reinforced concrete (RC) bridges do not employ seismic details and their structural performance has often been found to be inadequate under earthquake strong motions. In addition, existing seismic codes of practice, especially in Europe, do not contain specific guidelines for the reliable assessment of existing RC bridges. The present chapter illustrates the outcomes of two recent experimental seismic performance assessment carried out on typical portal frames and single columns existing reinforced concrete (RC) bridge designed for gravity loads only. The experimental response has been assessed by means of pseudo-dynamic and shake table testing. Intervention schemes employing innovative materials and technologies have also been considered as retrofitting measures to mitigate the onset of damage. The comprehensive experimental investigations prove the effectiveness of the isolation systems in preventing damage in the RC piers, especially limiting the maximum shear at the base of the piers, lowering the lateral drifts, preventing the onset of plastification in the frame sections and inhibiting the occurrence of the shear failure in the transverse beams of the RC portal frames of the piers.

Keywords Bridges · Seismic assessment · Seismic design · Existing structures · Experimental tests · Modelling

L. Di Sarno (✉)
Department of Engineering, University of Sannio, Benevento, Italy
e-mail: ldisarno@unisannio.it

F. Paolacci
Department of Engineering, University Roma Tre, Rome, Italy

© Springer International Publishing AG 2017
A.G. Sextos and G.D. Manolis (eds.), *Dynamic Response of Infrastructure to Environmentally Induced Loads*, Lecture Notes in Civil Engineering 2, DOI 10.1007/978-3-319-56136-3_6

6.1 Introduction

Widespread damage, partial and global collapses of bridges have frequently been surveyed in the aftermath of moderate-to-high magnitude earthquakes worldwide (e.g. Chang et al. 2000; Scawthorn 2000; FIB 2007; Li et al. 2008; Kawashima et al. 2010; among others). Most of the existing highway bridges, especially those employing reinforced concrete (RC) structures, were built during the 60s and 70s without seismic details; hence, their structural performance tends to be generally inadequate under earthquake ground motions (Shinozuka et al. 2000; Lupoi et al. 2003; Choi et al. 2004; Nielson and DesRoches 2007a, b; Pinto and Mancini 2009; Kappos et al. 2012). Brittle failure due to limited shear capacity is a common damage pattern experienced by non-ductile reinforced concrete (RC) bridges (e.g. Fig. 6.1), especially those employing either framed piers with short transverse beams or systems with short single piers. The failure could be caused by the inappropriate connection between piles and footing; it may be also due to an insufficient anchorage or lap splicing. Such damage pattern is exacerbated for RC bridge structures with smooth steel reinforcement, which is commonly found in existing RC bridges, especially those that were designed primarily for gravity loads only.

To date, numerous retrofitting strategies for existing deficient RC highway bridge structures have been proposed and validated through experimental testing and comprehensive numerical simulations (e.g. Priestley et al. 1994; Kawashima 2000; FHWA 2006; Wright 2011; among others). Such seismic retrofitting scheme include the use of innovative materials and/or technologies, such fiber reinforced-plastic, base isolation systems and/or supplemental damping. Base isolation system is a mature technique that can be cost-effective to protect existing

Fig. 6.1 Typical damage experienced by existing reinforced concrete bridges after an earthquake: sliding shear at *top columns* of the Cypress viaduct in the 1989 Loma Prieta earthquake (after NISEE, 2000)

deficient RC highway bridges (e.g. Priestley et al. 2007; Christopoulos and Filia-trault 2006; Matsagar and Jangid 2008; Attard and Dhiradhamvit 2009; Padgett et al. 2009, among many others). Applications of base isolation systems for railway bridges have also been recently investigated (e.g. Di Sarno 2013).

This chapter presents the outcomes of two recent experimental International research programs funded by the European Community, namely the RETRO and STRIT projects, carried out, the first on a typical RC frame piers highway and the second on a single columns RC bridge, both existing and designed for gravity loads only. The experimental response has been assessed by means of pseudo-dynamic (PsD) testing in the RETRO project and shake table (ST) testing in the STRIT project. The PsD experimental test campaign was performed at ELSA Laboratory of JRC (Ispra, Italy) within the Framework of the Seismic Engineering Research Infrastructures for European Synergies (SERIES), funded by the European Community in the 7th Framework Programme (Paolacci et al. 2014). The ST tests were carried out at the University of Naples, Federico II and were funded by the Italian Ministry of Research through the Project STRIT which deals with "Tools and Technologies for the Management of the Transportation Infrastructures".

In the RETRO and STRIT projects, intervention schemes employing innovative materials and technologies have also been considered as retrofitting measures to mitigate the onset of damage, especially to the transverse beams and the bridge piers. Friction pendulum bearings (FPBs) and high-damping rubber bearings (HDRBs) were utilized as isolation devices for the RC bridge continuous deck. Metallic strips and wires were also used to strengthen the critical sections of the RC bridge piers; the seismic performance of the retrofitted system has been assessed experimentally through incremental shaking table tests.

The comprehensive experimental PsD and ST investigations conducted on the sample bridge systems proved the effectiveness of the isolation systems in preventing the onset of damage in the RC piers, especially limiting the maximum shear at the base of the piers, lowering the lateral drifts, preventing the onset of plastification in the frame sections and inhibiting the occurrence of the shear failure in the transverse beams of the RC portal frames of the piers. Local interventions, comprising metallic strips and wires and aimed at strengthening the damaged critical sections of the RC piers, were found highly efficient to augment the member capacity and ductility.

6.2 Pseudo-dynamic Testing

6.2.1 The Bridge Sample Structure

The bridge used as sample structure is an existing reinforced concrete (RC) viaduct built in the 1970s and located in North-East of Italy, close to the location of the 2012 Emilia Romagna ($M_W = 5.9$) earthquake. The viaduct consists of a

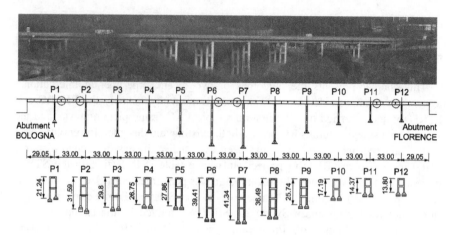

Fig. 6.2 The layout of the sample bridge

thirteen-span bay deck with two independent roadways resting on pairs of 12 portal frame piers, as displayed in Fig. 6.2. The portal frames comprise solid or hollow circular columns with diameters varying between 1.2 and 1.6 m, connected at the top by a cap-beam and at various heights by one or more transverse beams of rectangular sections. The total length is equal to 421.10 m. The shortest span is 29.05 m long and the longest is 33 m. The minimum and maximum pier heights are 13.80 m (P12) and 41.34 m (P7), respectively. Figure 6.2 provides the bird-view and the structural layout of the sample bridge.

The viaduct includes six intermediate Gerber saddles, acting as expansion joints; their location is indicated with circles in Fig. 6.2. Transverse RC diaphragms exist at the supports and also at intermediate locations in the span. The intermediate diaphragms are 3 in the first and last span close to the abutments; they are 4 in the remaining bridge spans.

The deck has an open cross section in the middle-span, as shown pictorially in Fig. 6.3, the latter figure also shows typical RC portal frames of the Rio Torto bridge.

The top slab width and thickness are 10.75 m and 0.20 m, respectively; the two girder webs are 0.35 m thick; the total depth is 2.75 m. The cross-section of deck close to the supports is a box girder due to the addition of a bottom RC slab. The bridge was designed with two decks for each road lane. Therefore, the system comprises two independent bridges for both gravity and horizontal loads. The bridge foundations include large RC block resting on rock, hence soil-structure interaction can be neglected in the earthquake response assessment as well as uplifting phenomena. The uniformly distributed weight of the deck is approximately 170 kN/m for each roadway.

The bridge piers have two types of transverse section: a solid circular section with diameter of 1200 mm and an hollow section with external and internal diameters equal to 1600 mm and 1000 mm respectively. The solid section has bars

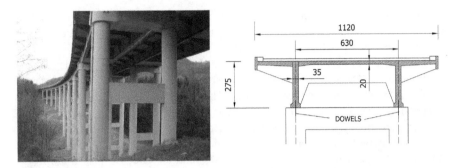

Fig. 6.3 Perspective view of the Rio Torto portal frames (*left*) and transverse section of the deck (*right*)

φ 20 mm whereas the hollow section presents φ 20 and φ 16 mm steel bars, external and internal respectively. The transversal reinforcement comprises a φ 6 mm steel spiral with a spacing of 14 mm. The transverse beams have a rectangular section with a width of 400 mm; the height varies between 1200 mm and 1500 mm. The longitudinal reinforcement include φ 24 and φ 20 steel bars. The transversal reinforcement comprises φ 8 steel bars with spacing of 200 mm and 45° inclined bars. The cap-beam of all the piers presents a U-shaped section. For such a beam, φ 18 longitudinal steel bars are used, whereas the transversal reinforcement includes φ 8 mm steel bars. Each pier has a vertical load ranging between 4900 and 5600 kN, being the length of the bays between 29 and 33 m. The structural steel of the smooth reinforcement bars has a mean strength of 350 MPa; the mean compressive resistance of the concrete is 26 MPa.

For the isolated structure, it was decided to remove the Gerber saddles, i.e. any intermediate longitudinal and transverse joint, thus transforming the bridge into a continuous girder system. Two spherical friction pendulum bearings (FPBs) were installed on the top of each pier, one under each girder web. The latter devices were designed in compliance with modern displacement-based methods; further details on the design method can be found in Della Corte et al. (2013).

6.2.2 The Pier Prototypes

The laboratory tests carried out within RETRO project include two RC portal frame piers of the Rio-Torto viaduct, as illustrated in the previous paragraph; such piers were tested by using PsD techniques with sub-structuring. The specimens are 1:2.5 scaled models of piers 9 and 11 of the sample bridge configuration. Pier 11 is a 2 floor 1-bay portal frame; the frame total height is H = 700 cm. The foundation is a rigid RC 600 cm long and 280 m width block. The columns of the portal frame specimen employ solid circular sections with diameter D = 48 cm. The longitudinal steel reinforcement includes 20 plain bars φ 8 mm and spiral smooth stirrups φ 3 with a spacing of 5.6 cm. Pier 9 specimen consists of a 3 floor 1-bay frame with a total height H = 1150 cm from ground. Further details of the sample pier specimens are provided in Paolacci et al. (2014) and Abbiati et al. (2015).

The portal frame pier specimens were tested in the as-built and seismic isolated configurations. The isolators were sub-structured and connected to the foundation blocks of the piers as described hereafter. The design mechanical properties of the isolation devices include a curvature radius R equal to 3000 mm and a nominal friction coefficient μ_f equal to 4%. Considering the scale factor 1:2.5 employed for the experimental tests, the prototype properties are R and μ_f equal to 1200 mm and 4%, respectively. Thus, the equivalent stiffness and the effective period are K_{eff} = 0.672 kN/m and T_{eff} = 1.63 s, respectively.

6.3 Test Set-up

The test set-up employed for the sample two scaled piers in the non-isolated configuration is shown in Fig. 6.4. The base of the specimens is fixed to the reaction floor by means of 16–36 mm diameter pre-stressed vertical Dywidag bars. To prevent cracking and excessive deformation of the base during testing, 16 and 10 φ 36 mm Dywidag bars are pre-stressed in the tangent and normal directions of testing, respectively. The two specimens are placed close to the rigid RC reaction wall of ELSA laboratory as shown in Fig. 6.4. The steel rig consists of HEB and C-shaped steel sections. Stiff rigid plates are used to connect the rig to the cap beams in each pier.

Fig. 6.4 Test set-up: global view of the piers close to the reaction walls (*left*) and FPBs test set-up (*right*)

Two 500 kN horizontal actuators with a displacement capacity of ±50 cm are connected to the steel rig, with a lever arm of 0.8 m with respect to the top of the cap beam. Such lever arm accounts for the distance from the centre of mass of the deck to the top of the cap beam of the prototype. The horizontal actuators are displacement-controlled so that during testing the rotation in the horizontal plan is inhibited. Additionally, a metallic system to restrain unexpected out-of-plane displacements of the piers was installed for safety requirements. Such system did not interact with the direction of testing. Vertical gravity loads are applied to the piers by means of two 500 kN capacity vertical jacks. The latter jacks are connected to a 36 mm Dywidag bar running through the centre of the columns and connected to the base by means of a nut (the vertical load is self-equilibrated).

The test set-up consisted of 18 actuators. Each actuator has a displacement transducer, measuring the displacement of the piston with respect to the cylinder, a load cell, measuring its applied axial force and some other sensors for the oil pressure at the chambers and for the servo-valve spool displacement. For this set-up, the horizontal actuators also had a displacement transducer measuring the displacement at a point of the specimen with respect to a fixed reference frame. These measurements, as well as the reference target and the servo-valve command are dealt, for each actuator, by a slave controller. A number of up to four slave controllers are connected to a master controller that exchanges all signals at the controller.

Linear Variable Differential Transformers (LVDTs) were installed in order to measure local deformation (e.g. curvatures) and global displacements. For example, three wire sensors were installed on Pier 9 (tall pier) to measure global displacement at each level of the transverse beams, together with 74 sensors installed in different levels above the column base, lower and upper beam-column joints and in the transverse beams. On each transverse beam, a triangular lattices of LVDT transducers were set to capture cross-section responses of transverse beams and joints. A similar sensor layout was also utilized for Pier 11 (short pier).

The PsD tests on the sample bridge retrofitted with seismic isolation system located at the base of the viaduct deck are carried out by means of the sub-structuring technique. The isolation system consists of 4 friction pendulum devices; during the tests, such devices are located at the ground level. They are connected to the pier specimen foundation blocks as shown in Fig. 6.4.

Two friction pendulum devices, with the sliding surfaces pointing upwards, are connected to the above steel plate. An additional steel plate, supported on the above friction isolators, is connected to the horizontal actuator, which can display horizontally along the transversal bridge direction. A set of two isolators with sliding surfaces pointing downwards are then placed on the top face of the latter steel plate. A third horizontal steel plate is placed on the second set of two isolators; the latter plate is loaded by the axial forces generated by the four vertical actuators. The plan layout of the set-up used for testing the FPBs is shown in Fig. 6.5.

For kinematic compatibility, it is essential to impose the continuity of displacements at the interface between the isolation system and the piers, for the same level of vertical loads.

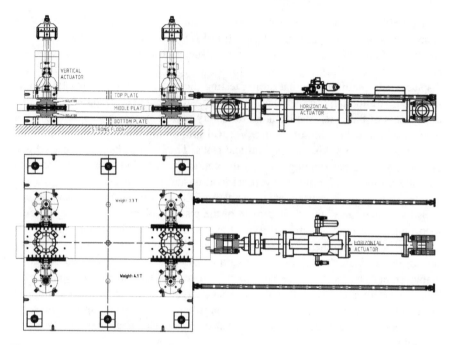

Fig. 6.5 Layout of the FPBs test set-up: cross section (*top*) and plan view (*bottom*)

6.4 Experimental Tests

The PsD testing method combines the numerical time integration of the equations of motion of a structure, properly condensed in a limited number of degrees-of-freedom (DoF), and the experimental measurement of the reaction forces resulting from this motion, applied by means of actuators. The hybrid analytical/experimental component is taken into account when introducing the sub-structuring technique (Dermitzakis and Mahin 1985), thus obtaining the dynamic response of a structure with only a part, usually the most vulnerable one, present in the laboratory. At ELSA Laboratory in Ispra, they have been using the so-called Continuous PsD scheme for many years (Magonette 2001). This scheme allows to load the experimental structure continuously (elapse time 2 ms) by performing in the same loop the time integration of the PsD model of the structure and the digital control of the actuators loading the structure. By avoiding the hold period associated with standard PsD implementation, the continuous method avoids load relaxation problems, optimizes the ratio signal/noise associated with the experimental errors, works with a constant time dilation and thus globally improves the quality of the results.

The test of the Rio Torto viaduct is substructured, thus a subassemblage of the structure is in the laboratory (experimental structure), and the remaining part is modeled numerically (numerical structure). The RETRO project includes large scale specimens, some of them (the piers) accumulating damage. It was thus needed to substantially upgrade the substructuring implementation traditionally used in ELSA, in order to have during the tests the same standards for error and alarm management, the same input (plus the additional requested information specific of substructuring), the same output definition (adding few substructuring related variables) as what is the current state-of-the art in continuous PsD testing at ELSA to handle conveniently and safely large-scale structures. In so doing, refined and simplified finite element (FE) models were implemented to perform the PsD in an efficient manner, without losing accuracy in the response analysis. The numerical models were implemented in the non-linear code OpenSees (McKenna et al. 2007); further details on the FE models used for the PsD tests can be found in Abbiati et al. (2015).

The Rio Torto viaduct is located in the Emilia Romagna region, North-East of Italy. The expected peak ground acceleration (PGA) ranges between 0.23 and 0.25 g, whereas for the collapse prevention condition (probability of 2% in 50 years) PGA ranges between 0.30 and 0.35 g. Given the geographical position of the bridge and the recent earthquake swarms occurred in the region (especially the earthquake records of the 20th and 29th May 2012), the seismic records of the 2012 Emilia (Italy) earthquakes were utilized for the PsD testing. The Mirandola records (MRN station) were utilized because of their seismological characteristics, i.e. PGAs and duration of the accelerograms, match the outcomes of the shake maps for the Emilia Romagna. The record of May 29th East-West was used for the

Table 6.1 PsD test program

Label	Configuration	Physical substructures	Accelerogram and PGA scaling
k06	*as built*	Piers #9 and #11	SLS, 10%
k07	*as built*	Piers #9 and #11	SLS, 100%
l01	*isolated*	Piers #9 and #11	SLS, 100%
l02	*isolated*	Piers #9 and #11	ULS, 100%
n01	*isolated*	FPBs #9 and #11	SLS, 100%
p01	*isolated*	Piers #9 and #11 & FPBs #9 and #11	SLS, 100%
p02	*isolated*	Piers #9 and #11 & FPBs #9 and #11	ULS, 70%
q01	*isolated*	Pier #9 & FPB #9	SLS, 100%
q02	*isolated*	Pier #9 & FPB #9	ULS, 65%
q03	*isolated*	Pier #9 & FPB #9	ULS, 65%
k09	*as built*	Piers #9 and #11	ULS, 100%
k10	*as built*	Piers #9 and #11	ULS, 100%
k12	*as built*	Piers #9 and #11	ULS, 200%
r01	*isolated*	FPB #9	ULS, 65%
r02	*isolated*	FPB #9	ULS, 80%
r03	*isolated*	FPB #9	ULS, 90%

98   L. Di Sarno and F. Paolacci

serviceability limit state (SLS) and the North-South component was used to assess
the seismic performance at the ultimate limit state (ULS).

The experimental program included two sets of tests: the first group relates to the
assessment of the seismic behaviour of the Rio-Torto viaduct; the second group of
tests deals with the evaluation of the effectiveness of FPBs to augment the seismic
performance of the retrofitted bridge system. The testing sequence includes the
experiments summarized in Table 6.1.

6.5   Experimental Results

6.5.1   Test Results on the Non-isolated Viaduct

A number of tests were carried out on the single piers to determine the initial lateral
stiffness and to check the earthquake response for low-intensity earthquakes. When
a 100% strong motion at SLS is considered, i.e. for test k07, the force-deflection
cycles show that the global behaviour is elastic. Minor hairline cracks were detected
at the transverse beams in pier 11 and 9 due to shear damage. The drift corresponding to a slight damage was about 3 cm for the tall pier and 1.6 cm for the
short pier; the drift ratio being about 0.3%. During the PsD test, the short pier
reached a displacement of about 3 cm, which corresponds to a drift ratio of about
0.6%. Such drift was mainly due to the higher horizontal deformability of pier 11,
with respect to the designed lateral flexural stiffness.

A significant shear crack pattern was observed in the transverse beams of both
piers, when subjected to 100% of ULS, i.e. during the k09 test. Few cracks were
observed between the cap beam and column joints. Base-column openings also
occurred during the lateral loading. Figure 6.6 shows the force-deflection response
computed during the k09 tests; high non-linearity can be identified in both the
physical piers.

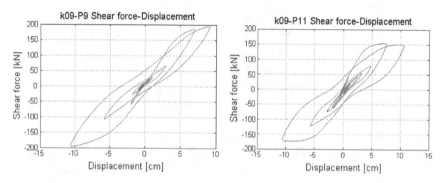

Fig. 6.6 Force-deflection cycle of pier 9 (*left*) and pier 11 (*right*) for test k09 (ULS PGA = 100%)

Fig. 6.7 Shear cracks pattern in the transverse beam (*left*) an column (*right*) of pier 11 after the test *k09* (ULS PGA = 100%)

The high level of shear damage in the beams and the large amount of crack opening at the column base and top is also clearly shown in Fig. 6.7.

To assess the effects of aftershocks on a damaged bridge, the input loading used for test *k09* was repeated with same sequence (test *k10*). The test aimed at verifying the level of degradation of non-linear behaviour and the level of damage of the both the piers. The test results show a decrease of the stiffness due to the increasing of the fix-end rotation effect, given by the high slippage of the reinforcing bars at the top and the bottom section of columns. This effect was magnified by the increased openings of shear cracks in the transverse beams.

To test the piers beyond-design conditions and quantify the global failure in terms of local and global response, a test with PGA = 0.54 g, i.e. 200% ULS, was also performed (test *k12*). A 2.4% drift ratio was reached in the short pier, thus the shear failure occurred in the transverse beams. Extensive damage occurred in both physical piers 9 and 11, as shown in Fig. 6.8.

However, the damage is widespread in the short pier, especially in the transverse beam, where large typical shear cracks were found. A large zone affected by cover spalling and buckling of the steel longitudinal bars was observed in the transverse beam and at the base column of short pier as per Fig. 6.8.

Similar damage failure was detected in the transverse beam at first floor of the tall pier; the damage in such slender pier is not widespread. The total drift ratio in pier 9 is 1.2%, which may cause severe damage in the transverse beam but it is far from the ultimate condition. Steel bar buckling occurred also at the base of the pier.

6.5.2   Test Results on the Isolated Viaduct

Comprehensive PsD tests were carried out on the isolated viaduct at SLS and ULS. The value of the friction coefficient (μ) used during the test is 4%. The bridge deck

Fig. 6.8 Crack pattern in the transverse beam (*top left*) and base column (*bottom left*) of the tall pier and buckling of the transverse beam (*top right*) and base column (*bottom right*) after k12 test (PGA = 200% ULS)

was assumed continuous, i.e. the saddles were eliminated, when isolation system was introduced at the base of the RC deck.

The bridge framed piers behave elastically when subjected to the 100% SLS earthquake input. The effectiveness of the FPBs in protecting the bridge is demonstrated by the nearly linear behaviour of physical as well as numerical piers at SLS. Transverse beams also exhibited an elastic response at SLS; the shear deformations in the lower transverse beam of pier 11 during test l01 vary linearly. The linear elastic response of the piers stems also from the curvature at the bottom section of left column of pier 11 during test l01. The hairline cracks on the transverse beams and at the bottom section of columns which appeared in the non-isolated configuration, namely in test k07, were not detected when base isolation devices were employed, i.e. for test l01.

The bridge configuration comprising physical piers (9 and 11) and physical isolators for piers 9 and 11 with $\mu = 4\%$ was tested with a PGA equal to 70% ULS. The retrofitting system, i.e. base isolation was utilized, for a bridge possessing slight damage. The response is nearly elastic, as shown in Fig. 6.9, where the force-deformation curves are plotted for the short and tall piers, respectively.

The effectiveness of the physical FPBs in protecting the bridge was further demonstrated when the strong motion at ULS is considered. T*ests r01, r02, r03,*

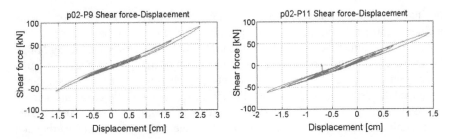

Fig. 6.9 Force-deflection cycle of Pier 9 (*left*) and Pier 11 (*right*) during test p02 (ULS PGA = 70%)

which correspond to acceleration equal to 65, 80 and 90% of ULS, i.e. 0.25 g (see also Table 6.1). For such tests, the physical components of the sample model comprise pier 9 and the isolators connected to that pier; all remaining piers and isolators are simulated numerically. This is a typical sub-structuring approach, utilized to estimate the response of the physical pier and devices.

6.6 Shake Table Testing

6.6.1 The Bridge Sample Structure

The bridge layout used as benchmark for the sample tested sub-assemblage comprises an existing RC multi-span simply supported bridge system, with two highway traffic lanes (total width of 7 m) and 22.65 m long girders. The deck consists of three RC pre-stressed precast 1.25 m high I-beams with 0.25 m thick solid deck. The deck is simply supported to represent typical configurations of existing RC bridges systems in the Italian highway network. The cantilever piers are RC solid circular sections with a 1.80 m diameter; the clear height of the cantilever member is 3.90 m with a 7.00 m wide and 1.50 m thick cap beam. The piers are fixed at the base to a 1.8 m thick RC foundation block. It is assumed that the bridge is resting on very dense soil, hence the soil-structure interaction can be neglected. The shear span of the reference bridge piers is short (aspect ratio is 2.60 = 3.90 m/1.80 m); thus, it is expected that the shear behaviour may significantly affect the member seismic response. The longitudinal reinforcement, which is arranged in a single layer and is uniformly distributed along the perimeter of the circular section, consists of 30 mm diameter bars, with a volumetric ratio, ρ_l, about 0.70%. The transverse reinforcement comprises a 12 mm diameter continuous spiral with a 250 mm spacing, thus ensuring the lower limit for the typical ρ_v—values of existing piers in the Mediterranean Region. The concrete cover for the reference piers is about 30 mm. The axial load ratio adopted in the piers is 0.05. The concrete compressive strength designed for the tests ranges between 20 and 25 MPa, which corresponds to a material that has experienced ageing degradation due to the harsh in situ exposure environment. The longitudinal and transverse steel reinforcements include Aq50 grade (mean yield strength of 370 MPa) smooth bars.

It is assumed that the reference multi-span simply supported RC highway bridge is located close to the city of Avellino, in the South of Italy, where historical faults have been identified (region with moderate-to-high seismic hazard). Such faults generated the devastating 23rd November 1980 ($M_S = 6.9$) earthquake, also known as Irpinia earthquake, which was characterized by three distinct sub-events occurring within the 40 s, along different faults. The selection of the Irpinia earthquake swarm allowed the evaluation of the effects of the cumulative structural damage on the tested bridge.

6.6.2 The Bridge Prototype

A scale factor of 1:3 for the length was used to reduce the total mass of the system. The peak ground accelerations (PGAs) were not scaled to have realistic dynamic forces without additional masses. This is also applicable to the elastic moduli which cannot be easily modified in the structural materials. For dynamic systems, once the scaling factors of 3 independent parameters are fixed, other properties of the system can be calculated from dimensional analysis (e.g. Moncarz and Krawinkler 1981), as outlined in Table 6.2.

The use of a small scaling factor for member length led to structural member dimensions and material properties compatible with those adopted in the formulations and calibrations of analytical models for global member capacity. To

Table 6.2 Scaling factors for the tested prototype

Quantity	Scaling law	Scale factor
Length	L_r	0.33
Acceleration	a_r	1.00
Modulus of elasticity	E_r	1.00
Time	$L_r^{1/2}$	0.57
Frequency	$L_r^{-1/2}$	1.73
Velocity	$L_r^{1/2}$	0.57
Displacement	L_r	0.33
Area	L_r^2	0.11
Mass	$E_r L_r^2$	0.11
Rotational mass	$E_r L_r^4$	0.01
Force	$E_r L_r^2$	0.11
Stiffness	L_r	0.33
Moment	$E_r L_r^3$	0.04
Energy	$E_r L_r^3$	0.04
Strain	ε_r	1.00
Stress	E_r	1.00

Fig. 6.10 Perspective of the test set-up (*left*), hinge and roller used as support devices (*right*) for the as-built configuration

reproduce realistic bond forces, both interior reinforcements and concrete aggregate were scaled considering the product availability and mechanical characteristics. The use of micro-concrete was not considered to avoid significant modification in the material properties. Longitudinal reinforcement bars with 10 mm in diameter and maximum aggregate size of 13 mm were used in the prototype construction.

The scaled prototype is a 7.55 m span bridge, consisting of two piers and a simply supported bridge deck, as depicted in Fig. 6.10. The deck supports consist of a steel cylindrical hinge (pinned condition) and a Teflon-steel slider (roller). The bridge deck, which has a total mass of about 15.40 tons, was primarily utilized to simulate the inertial forces during the shake table induced motions. The total pier height is about 1.80 m, with a cross-section diameter of 0.60 m. Further details on the prototype dimensions are provided in Fig. 6.11. Smooth bars were used for internal longitudinal and transverse reinforcements.

For the piers, 25 longitudinal 10 mm-diameter bars anchored with end-hooks and a continuous spiral 6 mm diameter, 100 mm spaced, were used for the longitudinal and transverse reinforcements, respectively. The average compressive strength of concrete is $f_{cm} = 22.5$ MPa, for the bridge pier and the pier cap and $f_{cm} = 33.5$ MPa, for the foundation block. The piers were subjected to a constant axial load ($v = P/A_g f_{cm} = 0.05$), as for the reference multi-span bridge system which was illustrated earlier. To achieve the target axial load, an additional force equal to 240 kN, was introduced at top pier by mean of a pre-stressed high-tensile strength steel rod.

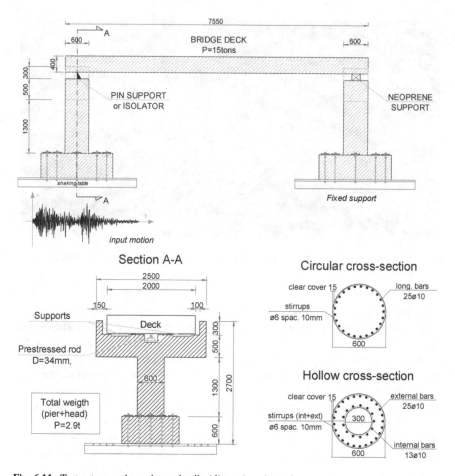

Fig. 6.11 Test set-up and specimen details (*dimensions in mm*)

6.6.3   The Test Set-up

The specimen was monitored with strain gauges, laser devices and tri-axial accelerometers, to measure internal deformations, local displacements and absolute accelerations during dynamic testing. Linear displacement sensors, i.e. LVDTs, with a 20, 50 and 100 mm stroke were installed to monitor the rotation at the base of the pier. The LVDTs were mounted vertically on steel rods cast horizontally in the column in the East-West. The rods were placed parallel and were located as close as possible to the expected concentration of inelasticity at the base of the pier. The LVDTs were employed to evaluate the fix-end rotation at the pier base. Electrical resistance strain gauges were symmetrically installed on four reinforcing bars at both East and West faces of the column to monitor axial strains.

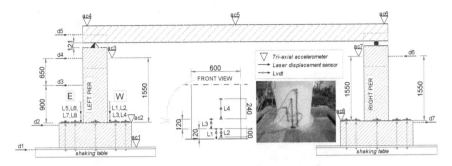

Fig. 6.12 Layout of the instrumentation installed on the test specimen (*dimensions in mm*)

Tri-axial accelerometers were installed to monitor the dynamic response of the test specimen. These sensors were used to measure vertical and horizontal, both longitudinal and transversal, accelerations. The layout of the instrumentation installed on the sample bridge system is shown in Fig. 6.12.

The data recorded by installed sensors enabled the computation of the inertial forces generated on the specimen from the base excitation. Laser Triangulation Displacement Sensors (LTDS) were used to evaluate displacements along the piers height and drift between piers top and the deck in the longitudinal and transversal directions.

6.6.4 Testing Program

The experimental tests carried out on the bridge prototype included more than 250 shakings, at increasing PGA; both unidirectional and bidirectional inputs were considered. Random vibrations were also utilized to detect the occurrence of the structural damage and hence the variation of the lateral stiffness of the prototype system.

Different layouts of the simply supported bridge deck were assessed: (i) with roller and cylindrical hinge, (ii) with rubber pads, (iii) with friction pendulum isolators and (iv) with high-damping rubber bearings (HDRBs). The damaged RC piers were also retrofitted, at the base sections, with novel fibre reinforced wrapping and wires. The retrofitted system was then tested in the configuration with roller and cylindrical hinge. The shake table testing program has been outlined in Table 6.3.

The PGA scaling was not uniform for all tests as different failure modes were detected during the shake table testing. For the configuration with the isolators the maximum PGAs that was applied to the table was controlled by the allowable horizontal displacements of the FP devices ad HDRBs. When using the rubber pad, the maximum PGA was controlled by the sliding of the supports.

The East-West and North-South (horizontal) components of the strong motion recorded at the Calitri station (1980 November 23 Irpinia (Italy) $M_S = 6.9$

Table 6.3 Shake table testing program

Label	Configuration	Support condition	Pier type	Earthquake input	PGA scaling
I01	isolated	Friction Pendulum–Friction Pendulum	Solid circular section	Unidirectional Calitri E-W	(5-10-25-50-75-100-150-200-250-300-350-400)%
I02	isolated	HDRB–HDRB	Solid circular section	Unidirectional Calitri E-W	(5-10-25-50-75-100-150-200-250-300-350-400)%
E01	as-built	Rubber Pad–Rubber Pad	Solid circular section	Unidirectional Calitri E-W	(5-10-25-50-75-100-150-200)%
E02	as-built	Roller–Cylindrical Hinge	Solid circular section	Unidirectional Calitri E-W	(5-10-25-50-75-100-150-200-250-300-350-400)%
R01	retrofitted	Roller–Cylindrical Hinge	Solid circular section	Unidirectional Calitri E-W	(5-10-25-50-75-100-150-200-250)%
I03	isolated	Friction Pendulum–Friction Pendulum	Hollow circular section	Bidirectional Calitri E-W & N-S	(5-10-25-50-75-100-125-130)%
E03	as-built	Roller–Cylindrical Hinge	Hollow circular section	Bidirectional Calitri E-W & N-S	(5-10-25-50-75-100-125)%

earthquake), which is located closer than 20 km to the fault, were selected as input motions for the shake table tests on the sample specimen. The PGA of the as-recorded Calitri strong motion is 0.17 g along both horizontal directions. The total duration of the as-recorded strong motion is 85 s. During the unidirectional tests, the Calitri record was scaled from low-intensity shaking to the bridge column collapse. The selected earthquake record included the effects of multiple events thus it was utilized to evaluate the accumulation of damage, if any, on the sample bridge system.

The test protocol included the Calitri ground motions with white-noise excitation between them. The latter were utilized to estimate the period elongations, if any, which were caused by the damage occurrence.

6.6.5 Discussion of the Experimental Results

The prototypes systems with circular solid and hollow section RC piers were subjected to a series of time-histories at increasing PGA, as illustrated in the previous paragraph (see also Table 6.2). The effects of cumulative damage were investigated especially on the as-built configurations, where damage occurred in the base sections of the piers when using the simply support configuration with cylindrical hinge and roller. The results of the as-built system are provided below along with the outcomes of the response for the retrofitted system; emphasis is on the circular solid piers. It can be assumed that the bridge deck is rigid in the axial direction as the differences in the recorded accelerations along its length are negligible.

6.6.5.1 Test Results on the as-Built System

Modal response analysis of the prototype was initially carried out. The single RC solid pier exhibited a fundamental period equal to 0.034 s. Such value was determined by using the transfer function method and the white noise as input. The mean value of the equivalent viscous damping is 2%, which is significant lower than the values implemented in the current seismic standards, i.e. 5% (e.g. CEN 2006, among others).

The periods of the bridge prototype were computed for increasing acceleration amplitudes. It is observed that the prototype experienced a significant period elongation, i.e. from 0.13 to 0.36 s. The latter period elongation corresponds to significant damage in the bridge system in the configuration with hinge and roller at supports (tests E02). It was experimentally observed that the damage concentrated at the base of the bridge pier, as further discussed hereafter. The equivalent viscous damping coefficient estimated experimentally for the bridge prototype varies between 7 and 10%; the mean value is 8%. It can be stated that when system configurations are considered the exiting code formulations provide a conservative estimate of the equivalent viscous damping (underestimation of 60%, i.e. 5 vs. 8%).

In the configuration with rubber pad on both pier caps (tests E01), it is found that for values of PGA lower than 0.10 g, under unidirectional loading, the sliding of the deck does not occur. As the PGA increases, the bridge deck tends to display laterally. A linear correlation between the values of PGA, used as strong motion input for the ST, and the maximum sliding of the deck was determined. For PGA = 0.15 g, the maximum lateral displacement measured during the ST test was 15 mm. At PGA = 0.35 g, i.e. for 200% scaling factor, the lateral displacement was 50 mm (which corresponds to 150 mm in the full scale). The tests were interrupted at 0.35 g, as it was considered that for larger PGA the maximum sliding of the bridge deck may cause hammer between adjacent beams. Neoprene pads have been installed during the years on numerous bridges in Italy (e.g. Borzi et al. 2015); it can be assumed that such pads may dissipate large earthquake-induced energy through the sliding of the deck, thus protecting the RC piers from widespread damage and the structural collapse. The occurrence of the sliding limits the maximum forces that can be transmitted to the piers and the foundation systems.

For the simply supported configuration where roller and cylindrical hinges are used (tests E02), the system behaves elastically for PGA lower than 0.21 g; the base acceleration is slightly amplified at the bridge deck level. Increasing the ground motion intensity, the maximum recorded acceleration at the top of the deck is significantly higher than the PGA used as input signal for the ST. Such response is caused by the dynamic excitation of the bridge prototype and to the flexibility of the left pier that amplifies the base acceleration until the strong motion where the maximum amplification factor, i.e. 1.7 with respect to the base input, has been recorded. During the test with PGA = 0.64 g, the pier first cracking was detected. A marked crack at the base of the pier was detected; this crack pattern is generated by the construction joint and the lap splice of longitudinal smooth reinforcements, which create a flexural critical cross-section at the interface of the pier and the foundation. The foundation block employed higher compressive strength concrete, thus the observed crack affected primarily the bottom of the pier. Once the column first cracking occurred, at larger PGAs significant increase in the crack width were detected. However, no further cracks were observed along the pier height. This allowed the hinge mechanism development at the pier base consisting of longitudinal reinforcement yielding and concrete cover spalling. The pier exhibited flexural yielding for PGA = 0.85 g; such yielding acted as a fuse in the base acceleration propagation along the pier height. For larger PGAs, the maximum recorded top acceleration is lower than the PGA applied at the table and a relevant rocking mechanism was observed. The latter damage mechanism is typically found in RC columns with smooth longitudinal reinforcements and significant slip at the base of the structural member. Such slip results in a significant increase in the pier top displacement because of the member fixed-end rotation. During the rocking mechanism, the bridge prototype experienced sudden changes of the lateral stiffness, which jeopardized the control of the shaking table input. As a result, large

Fig. 6.13 Maximum pier
drift as function of the PGA
applied on the shake table

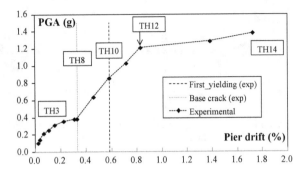

spikes in the shape of the input acceleration records were observed for the time
histories with PGAs higher than 0.64 g.

In the tests E02, the bridge deck exhibited minor sliding (about 20 mm for the
prototype) when the unidirectional 100% East-West Calitri strong motion was used
as input motion. Such sliding increased linearly for PGAs lower than 0.4 g (about
45 mm for the prototype); the displacement variation became highly nonlinear for
larger PGAs: at 0.7 g the maximum sliding is about 115 mm.

The lateral drifts for the pier of the prototype system with roller and cylindrical
hinge are plotted in Fig. 6.13 as function of the PGAs used as input for the shake
table.

The results depicted in Fig. 6.13 show that three changes occurred in the
experimental response. For PGA lower than 0.21 g, the pier drift increases almost
linearly with the values of acceleration input; thus, the pier behaves elastically. The
analysis of the local quantities shows that the shear response governs the pier
deformability. For increased PGAs (lower than 0.4 g), a faster increase in the drift
is observed. Such response is caused by the change in the pier stiffness generated by
the slip of the longitudinal bars and hairline cracking. The vertical deformations at
the column base become in the order of 1 mm and, in turn, the fixed-end reaches the
50–60% of the total drift. At a PGA = 0.4 g (which corresponds to a 0.33% lateral
drift) the first crack was detected at the pier base and a significant change of the
slope of the curve plotted in Fig. 6.13 was observed.

For accelerations larger than 0.4 g, high PGA increments produce changes in the
maximum pier drift but at a lower rate. Pier cracking affects the period elongation of
the RC member and, in turn, the reduction in the acceleration demand on the bridge
deck. For higher accelerations, the pier experienced the formation of the plastic
hinge at the base and the yield penetration along the member height. Furthermore,
after the yielding, the drift component related to the fixed-end rotation becomes
dominant respect to the other contributions, i.e. 75–80% of the total drift. The
flexural contribution is negligible with respect to the total drift demand. The distribution of the acceleration within the prototype at 0.6 g in the tests E02 is shown
in Fig. 6.14; the labeling of the accelerometers is provided in the Fig. 6.12. The
amplification at the top of the pier with respect to the input used at the foundation
block is negligible (about 10%), i.e. 0.67 g versus 0.60 g.

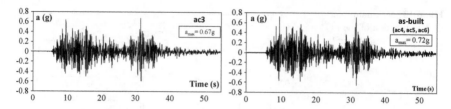

Fig. 6.14 Acceleration distribution along the pier height at PGA equal to 0.6 g in the as-built configuration (tests E02)

The analysis of the local measurement at the pier base points out that the position of the zero point of base deformations is almost constant for PGAs larger than 1.0 g and it is approximately 0.15D (with D the diameter of the pier cross-section) from the pier face. This confirms that a rocking mechanism characterizes the pier response rather than a pure flexural behaviour. A 3% maximum strain, which is close to the hardening strain, was recorded on the steel longitudinal bars. The latter strain value proves that the SLS of the RC pier was exceeded, but the response is far from the collapse limit state. At the end of the tests, a maximum pier drift about 1.7% was achieved and a notable damage consisting in concrete cover spalling and the beginning of bar buckling was detected. Although the pier possesses a low aspect ratio (L/h ≈ 2), no shear cracks were observed during the tests. Thus, it may be argued that existing squat RC members with smooth bars do not experience brittle shear mechanisms because of the fixed-end rotation and rocking of the pier. Additionally, the bond-slip mechanisms of smooth longitudinal bars in bridge piers limit the overstress in the foundation blocks because of the cutting of the base shear caused by the member rocking.

When the RC damaged pier was retrofitted with the use of metallic strip and wires applied at the critical section at the base of the pier, the ST tests showed that the system can withstand larger PGAs without experiencing any damage. Comparing the same level of base accelerations for the tests E02 and R01, e.g. 0.7 g, the retrofitted pier did not exhibit any crack pattern, thus proving the significant beneficial effects in terms of strength provided by the local intervention scheme comprising the use of novel materials. Such scheme can be very effective in the aftermath of moderate-to-large earthquakes to restore the transitability (functionality) of RC highway networks, especially those that essential to access critical facilities, e.g. hospitals, emergency centers, etc.

6.6.5.2 Test Results on the Deck-Isolated System

The shaking table tests carried out on the isolated configurations demonstrated the effectiveness of the tested bearings, e.g. FPBs and HDRBs, as global seismic retrofit solution. Such effectiveness was demonstrated experimentally for both

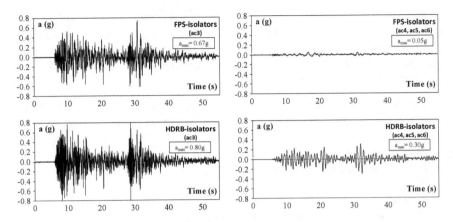

Fig. 6.15 Pier cap and deck accelerations in configuration with FPBs (tests I01) and HDRBs (tests I = 2) when using PGA 0.65 g

unidirectional and bidirectional earthquake loading at increasing PGAs applied during the ST tests carried out on the prototype systems, namely tests I01 and I02.

A significant reduction in the recorded accelerations can be observed comparing the acceleration at the top of the pier (ac3) with those transmitted on the deck (tests I01 and I02 shown in Fig. 6.15 for PGA = 0.65 g). As a consequence, a marked reduction on the shear demand was estimated on the RC piers, which exhibited an elastic response, even at large PGAs, for example higher than 1.0 g.

Significant differences can be observed in the devices effectiveness. The FPBs possess higher effectiveness than HDRBs in reducing the deck accelerations to very low values, as displayed in Fig. 6.15. Higher accelerations were transmitted by HDRB devices due to the higher lateral stiffness. Conversely, lower relative deck-to-pier displacements were observed using the HDRBs respect to the FPBs.

6.7   Conclusions

The experimental seismic response assessment of portal framed and single piers reinforced concrete (RC) existing highway bridges designed for gravity loads and located in the Mediterranean region have been discussed with respect to two recent experimental International research programs funded by the European Community. In the RETRO project pseudo-dynamic (PsD) testing was carried out at ELSA Laboratory of JRC (Ispra, Italy) on a 1:2.5 scaled model of short and tall RC portal frame piers, while shake table (ST) tests performed at the University of Naples, Federico II on a 1:3 bridge prototype were performed for the STRIT project.

As-built and retrofitted bridge system configurations were assessed in both experimental research programs. Local and global intervention strategies were investigated to mitigate the onset of damage, especially to the transverse beams and at the base of free standing single piers. Friction pendulum bearings (FPBs) and high-damping rubber bearings (HDRBs) were utilized as isolation devices for the RC bridge continuous deck. Metallic strips and wires were also utilized to strengthened the critical sections of the single columns RC bridge piers.

For the PsD tests on portal frame bridges, it is found that:

- Extensive damage pattern occurs, especially in the short physical pier, when the transverse beam has been subjected to a severe cracking damage due to shear beams;
- Significant fix-end-rotations occurred at the base of the pier columns. The latter rotations are caused by the high bond slip effect typically found for plain steel bars
- Significant slip of the longitudinal steel reinforcement, occur at the base of the bridge piers at about 0.40% lateral drift ratio;
- During tests performed at different amplitudes, frequency and pressure the friction coefficients of FPBs varies significantly. The actual friction values can be twice the nominal values utilized in the design of the seismic isolation devices.

For the ST tests, it can be observed that:

- The assessment of local and global response quantities shows that fixed end rotations, with significant slip of the longitudinal steel reinforcement, occur at the base of the bridge piers at about 0.40% lateral drift ratio;
- Piers with low aspect ratio (L/h $\approx$ 2), do not exhibit shear cracks. Thus, squat RC piers with smooth bars and low-strength concrete may not experience brittle shear mechanisms because of the fixed-end rotation and rocking of the structural member;
- Significant damage at the pier base, consisting in concrete cover spalling and the beginning of the steel longitudinal bar buckling, occurred at about 1.70% lateral drift. Additionally, at large drifts, residual strains were also identified in the longitudinal bars.
- When using local strengthening for the critical section at the base of the pier, the latter can withstand PGAs larger than 1.0 without experiencing any damage.

It can thus be concluded that the comprehensive experimental PsD and ST investigations conducted on the sample bridge systems emphasize the effectiveness of the isolation systems in preventing the onset of damage in the RC piers, especially limiting the maximum shear at the base of the piers, lowering the lateral drifts, preventing the onset of plastification in the frame sections and inhibiting the occurrence of the shear failure in the transverse beams of the RC portal frames of the piers. Local interventions, comprising metallic strips and wires and aimed at strengthening the damaged critical sections of the RC piers, were found highly efficient to augment the member capacity and ductility.

The above experimental findings are reliable and robust but it is firmly believed that further experimental tests and numerical simulations are needed to assess the seismic risk of existing RC bridges with low-strength concrete and smooth bars, especially when interactions between flexural and shear failure modes are expected. For such bridges, the effects of multiple earthquake ground motions may erode significantly the lateral stiffness resulting in large drift demand on the piers.

Acknowledgements This work was financially supported through the years by the European Community, with the following Projects of the 7th Framework Programme: Experimental & Computational Hybrid Assessment Network for Ground-Motion Excited Soil-Structure Interaction Systems (PIRSES-GA-2009-247567-EXCHANGE-SSI) grant (http://www.exchange-ssi.net); Assessment of the seismic vulnerability of an old RC viaduct with frame piers and study of the effectiveness of different isolation systems through pseudo-dynamic test on a large scale model (RETRO), funded by the "Seismic Engineering Research Infrastructures for European Synergies" (SERIES, FP7/2007-2013, GA-227887); "Tools and Technologies for the Management of the Transportation Infrastructures" (STRIT). Any opinions, findings and conclusions or recommendations expressed in this paper are those of the authors and do not necessarily reflect those of the funding agencies. The authors would like to acknowledge the contribution of ALGA Spa and FIP Industriale Spa, for having kindly provided the seismic isolation devices (FPBs and HDRBs) used for the pseudo-dynamic and shake table testings. The contribution of academic and technical staff at ELSA-JRC (Ispra) and Laboratory of the Department of Structures for Engineering and Architecture of the University of Naples Federico II is also highly appreciated.

References

Abbiati G, Bursi OS, Caperan P, Di Sarno L, Molina FJ, Paolacci F et al (2015) Hybrid simulation of a multi-span RC viaduct with plain bars and sliding bearings. Earthq Eng Struct Dyn. doi:10.1002/eqe.2580

Attard T, Dhiradhamvit K (2009) Application and design of lead-core base isolation for reducing structural demands in short stiff and tall steel buildings and highway bridges subjected to near-field ground motions. J Mech Mater Struct 4(5):799–817

Borzi B, Ceresa P, Franchin P, Noto F, Calvi GM, Pinto PE (2015) Seismic vulnerability of the italian roadway bridge stock. Earthq Spectra 31(4):2137–2161. doi:10.1193/070413EQS190M

CEN 2006. Eurocode 8—structures in seismic regions—design—part 2: bridges. Comitè e Europaèen de Normalisation: Brussels, 1994

Chang K-C, Chang D-W, Tsai M-H, Sung Y-C (2000) Seismic performance of highway bridges. Earthq Eng Eng Seismol 2(1):55–77

Choi E, DesRoches R, Nielson B (2004) Seismic fragility of typical bridges in moderate seismic zones. Eng Struct 26(2):187–199

Christopoulos C, Filiatrault A (2006) Principles of passive supplemental damping and seismic isolation. IUSS Press

Della Corte G, De Risi R, Di Sarno L (2013) Approximate method for transverse response analysis of partially isolated bridges. J Bridge Eng 18(11):1121–1130

Dermitzakis SN, Mahin SA (1985) Development of substructuring techniques for on-line computer controlled seismic performance testing. Technical report, Earthquake Engineering Research Center, University of California, Berkeley, CA, USA, 1985. Report No. UCB/EERC-85/04

Di Sarno L (2013) Base isolation of railway bridges. Int J Mech 1(7):302–309

Federation Internationale du Beton - fib - Task Group 7.4 (2007) Seismic bridge design and retrofit - structural solutions. Lausanne, Switzerland

FHWA, Federal Highway Administration (2006). Seismic retrofitting manual for highway structures: part I—bridges. Publication No.FHWA-HRT-06-032, Research, Development and Technology Tuner-Fairbank Highway Research Center, McLean, VA

Kappos AJ, Saiidi MS, Aydınoğlu MN, Isaković T (eds) (2012) Seismic design and assessment of bridges—inelastic methods of analysis and case studies. Geotechnical, geological and earthquake engineering, vol 21. Springer

Kawashima K (2000) Seismic design and retrofit of bridges. Bull New Zealand Soc Earthq Eng 33 (3):265–285

Kawashima K, Unjoh S, Hoshikuma J, Kosa K (2010) Damage of bridges due to the 2010 Maule, Chile earthquake. J Earthq Eng 15(7):1036–1068

Li J, Peng T, Xu Y (2008) Damage investigation of girder bridges under the Wenchuan earthquake and corresponding seismic design recommendations. Earthq Eng Eng Vib 7(4):337–344

Lupoi A, Franchin P, Schotanus M (2003) Seismic risk evaluation of RC bridge structures. Earthq Eng Struct Dyn 32(8):1275–1290

Magonette G (2001) Development and application of large-scale continuous pseudodynamic testing technique. Philos Trans R Soc Lond A 359:1771–1799

McKenna F, Fenves GL, Scott MH (2007) OpenSees: open system for earthquake engineering simulation. PEER, University of California, Berkeley, CA

Matsagar V, Jangid R (2008) Base isolation for seismic retrofitting of structures. Pract Period Struct Des Constr 13(4):175–185

Moncarz PD, Krawinkler H (1981) Theory and application of experimental model analysis in earthquake engineering, Stanford, CA, USA

Nielson BG, DesRoches R (2007a) Analytical seismic fragility curves for typical bridges in the Central and South-Eastern United States. Earthq Spectra 23(3):615–633

Nielson BG, DesRoches R (2007b) Seismic fragility methodology for highway bridges using a component level approach. Earthq Eng Struct Dyn 36(6):823–839

Padgett JE, DesRoches R (2009) Retrofitted bridge fragility analysis for typical classes of multispan bridges. Earthq Spectra 25(1):117–141

Paolacci F, Pegon P, Molina FJ, Poljansek M, Giannini R, Di Sarno L, Abbiati G, Mohamad A, Bursi OS, Taucer F, Ceravolo R, Zanotti Fragonara L, De Risi R, Sartori M, Alessandri S, Yenidogan C (2014) Assessment of the seismic vulnerability of an old RC viaduct with frame piers and study of the effectiveness of base isolation through PsD testing on a large scale model. JRC scientific and policy report. doi:10.2788/63472. ISBN 978-92-79-35271-3

Pinto PE, Mancini G (2009) Seismic assessment and retrofit of existing bridges, the state of earthquake engineering research in Italy: the ReLUIS-DPC 2005–2008 Project, 111–140, © 2009 Doppiavoce, Napoli, Italy

Priestley MJN, Seible F, Calvi GM (1994) Seismic assessment of existing bridges. Wiley Online Library

Priestley MJN, Calvi GM, Kowalsky MJ (2007) Displacement based seismic design of structures. IUSS Press, Pavia, Italy

Scawthorn C (2000)The Marmara Turkey Earthquake of August 17, 1999, Reconnaissance Report, MCEER Technical Report MCEER-00-0001, Buffalo, NY

Shinozuka M, Feng MQ, Lee J, Naganuma T (2000) Statistical analysis of fragility curves. J Eng Mech 126(12):1224–1231

Wright T, DesRoches R, Padgett JE (2011) Bridge seismic retrofitting practices in the Central and Southeastern United States. J Bridge Eng ASCE 16(1):82–92

Chapter 7
Modeling of High Damping Rubber Bearings

Athanasios A. Markou, Nicholas D. Oliveto and Anastasia Athanasiou

Abstract High damping rubber bearings have been used in the seismic isolation of buildings worldwide for almost 30 years now. After a brief introduction to the process leading to their manufacturing, a description is given of the main tests required by current seismic codes for the design of such devices. An extensive review is then presented of the models available in the literature for the simulation of the dynamic response of high damping rubber bearings under simultaneous shear and compression. Given the extremely complex and highly nonlinear behavior of these devices, no model is capable of capturing every single aspect of the dynamic response. Issues and uncertainties involved in the characterization of this complex behavior are pointed out. These include, among others, coupled bidirectional horizontal motion, coupling of vertical and horizontal motion, strength and stiffness degradation in cyclic loading, and variation in critical buckling load capacity due to lateral displacement. Finally, a novel 1D mechanical model for high damping rubber bearings is proposed, based on the combination of simple and well-known rheological models. The model is calibrated against a set of harmonic tests at strain amplitudes up to 200%. Extension to bidirectional horizontal motion and to varying vertical load is subject of ongoing work.

Keywords High damping rubber bearings · Mechanical models · Base isolation · Shear behavior · Constitutive modeling · Rubber

A.A. Markou (✉)
Norwegian Geotechnical Institute, Oslo, Norway
e-mail: athanasios.markou@ngi.no

N.D. Oliveto
Buffalo, USA
e-mail: noliveto@buffalo.edu

A. Athanasiou
Department of Civil Engineering and Architecture, University of Catania, Catania, Italy
e-mail: athanasiou@dica.unict.it

© Springer International Publishing AG 2017
A.G. Sextos and G.D. Manolis (eds.), *Dynamic Response of Infrastructure to Environmentally Induced Loads*, Lecture Notes in Civil Engineering 2,
DOI 10.1007/978-3-319-56136-3_7

7.1   Introduction

Natural rubber is obtained from a milky fluid (latex) extracted from the *Hevea Brasiliensis* tree, also called by the Maya Indians *Caoutchouc*, which means "weeping wood" (Treolar 1975). The name "rubber" was first used in 1770 by chemist Joseph Priestley, who observed that the material was very good for rubbing out pencil marks from paper (Treolar 1975). In 1939, the term "elastomer" was attributed by Fisher to synthetic materials having rubber-like properties. Natural rubber latex, as it comes out of the tree, consists of chains of polyisoprene, one of the most well-known polymers. The main disadvantage of the material is that it gets sticky when warm and brittle when it gets cold (Michalovic and Brust 2000). In 1839, Charles Goodyear discovered that the addition of sulfur to natural rubber latex, while heating, creates crosslinks between the chains of polyisoprene, forming a super-molecule. This process, called curing or vulcanization, generates the material that we nowadays call natural rubber. In 1931, DuPont invented the popular synthetic polymer Duprene, which was later called Neoprene (Ciesielski 1999). The material consists of chains of polychloroprene, crosslinked through vulcanization using metal oxides rather than sulfur. In order to improve their mechanical properties, fillers (mainly carbon black) are generally added to the rubber compound, accelerators are used to shorten the duration of heating during vulcanization, anti-ozonants are added to protect the material against ozone attack, and anti-oxidants are used to reduce ageing and to delay degradation due to exposure to oxygen (Constantinou et al. 2007).

In 1889, natural rubber pads were installed between the superstructure and the piers of a rail bridge in Melbourne, Australia. These were approximately 1.3 cm thick and were meant to absorb impact rather than to accommodate horizontal movement (Gent 2012). In 1954, French Engineer Eugene Freyssinet obtained a patent for his idea of reinforcing sheets of rubber with thin steel plates, see Fig. 7.1. By imposing steel plates between layers of rubber, a combination of vertical stiffness and horizontal flexibility was achieved (Kelly and Konstantinidis 2011). In 1956, a vulcanization procedure was adopted to bond the thin steel plates to the rubber sheets (Kelly and Konstantinidis 2011). Since then, multilayer rubber bearings have been used extensively in a variety of applications, one of the most popular being the protection of buildings against earthquakes through seismic or base isolation. Since rubber sheets provide very low damping, a lead plug is sometimes inserted in the bearings to increase energy dissipation (Naeim and Kelly 1999). These bearings are referred to as Lead Rubber Bearings (LRBs). An alternative way of increasing energy dissipation is to provide sufficient damping in the rubber sheets themselves by using fillers (Naeim and Kelly 1999). A high-damping natural rubber was achieved by the Malaysian Rubber Producer's Research Association (MRPRA) in 1982. Extra fine carbon black, oils or resins were used as fillers to increase damping of the compound. The first building using high-damping rubber bearings is the Foothill Communities Law and Justice Center, constructed in 1985 in the city of Rancho Cucamonga in Southern California.

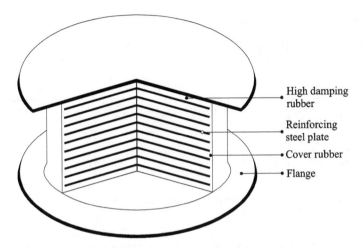

Fig. 7.1 High damping rubber bearing (HDRB) (from Bridgestone Catalog 2013)

In this study, both synthetic and natural rubber bearings, with damping capabilities enhanced by the methods described above, shall be referred to as High Damping Rubber Bearings (HDRBs). Rubber bearings that are not treated to increase their damping characteristics are referred to as Low Damping Rubber Bearings (LDRBs). Since these bearings do not exhibit creep, and can be modeled adequately as linear elastic systems with viscous damping, in this work they shall not be considered further. LRBs have high damping characteristics, and exhibit the well-known bilinear behavior obtained by combining a linear elastic LDRB and an elastic-perfectly plastic lead core (Bozorgnia and Bertero 2004).

HDRBs have a much more complex behavior than both LDRBs and LRBs. They exhibit high stiffness and damping at low shear strains, which minimizes the response under service and wind loads, and low shear stiffness, but yet adequate damping capacity, at the design displacement level. At higher displacement amplitudes they exhibit an increase in stiffness and damping, useful to limit displacements under major earthquakes. Additional important aspects ought to be considered when modeling the behavior of HDRBs in the design of seismic isolation systems. A major one is creep, which makes the behavior of the devices rate dependent. Mechanical properties are also affected by manufacturing variations, contact pressure, loading and strain history, temperature, and ageing (Constantinou et al. 2007).

7.2   Testing of High Damping Rubber Bearings

Seismic isolators are generally subjected to three kinds of tests (Shenton 1996; McVitty and Constantinou 2015): (i) qualification tests, (ii) prototype tests, and (iii) production tests.

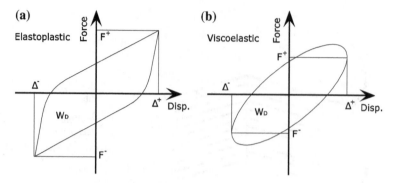

Fig. 7.2 **a** Elastoplastic and **b** viscoelastic hysteresis loops (from FEMA P-751 2009)

The qualification tests are carried out by the manufacturer, they are not project-specific and serve the purpose of characterizing the effects of heating due to cyclic dynamic motion, loading rate, scragging, variability and uncertainty in production bearing properties, temperature, ageing, environmental exposure, and contamination (McVitty and Constantinou 2015).

The prototype tests are project-specific and are conducted to verify the design properties of the isolation system prior to construction. They are performed on full-size specimens and consist of cyclic tests at different strain amplitudes and compression loads (Shenton 1996; McVitty and Constantinou 2015).

Production tests are also project-specific and are carried out to verify the quality of each individual isolator. All the project's isolators are tested, in combined compression and shear, and at not less than two-thirds of the maximum displacement (Shenton 1996; McVitty and Constantinou 2015).

The main mechanical properties of HDRBs, evaluated through cyclic tests under compression at different strain amplitudes, are:

(i) effective stiffness K_{eff}
(ii) effective shear modulus G_{eff}
(iii) equivalent viscous damping ζ_{eq}

The effective stiffness K_{eff} is defined as

$$K_{eff} = \frac{|F^+| + |F^-|}{|\Delta^+| + |\Delta^-|} \tag{7.1}$$

where, as shown in Fig. 7.2, Δ^+ and Δ^- are the maximum positive and minimum negative displacement amplitudes, F^+ and F^- the corresponding forces.

The effective shear modulus G_{eff} is given by

$$G_{eff} = \frac{K_{eff} T_r}{A_r} \tag{7.2}$$

where T_r is the total rubber thickness and A_r is the bonded rubber area.

Finally, the equivalent viscous damping ratio ζ_{eq} is defined as

$$\zeta_{eq} = \frac{2}{\pi}\left[\frac{W_D}{K_{eff}\left(|\Delta^+| + |\Delta^-|\right)^2}\right] \tag{7.3}$$

where W_D is the area enclosed by the hysteresis loop. Expressions 7.1–7.3 are based on the results of harmonic tests carried out in displacement control. These are not only amplitude dependent, but also frequency dependent. However, provided that the frequency is not too low, the dependence on frequency may be neglected. Ideally, $\Delta^+ = \Delta$ and $\Delta^- = -\Delta$ hold, but this is generally not the case due to measurement and implementation errors. The effective shear modulus, G_{eff}, provided by expression 7.2, relates the nominal shear stress, τ, to the nominal shear strain, γ. These are given by

$$\tau = \frac{F^+ - F^-}{2A_r} \tag{7.4}$$

$$\gamma = \frac{\Delta^+ - \Delta^-}{2T_r} \tag{7.5}$$

Clearly, the effective stiffness, K_{eff}, and equivalent viscous damping, ζ_{eq}, depend on the amplitude of displacement at which the bearings are subjected, and ultimately on the amplitude of the shear strain, γ. Moreover, these characteristics also depend on the vertical compressive load applied to the bearings, both in the test and during their service life.

Figure 7.3 shows the third cycle force-displacement responses of displacement controlled harmonic tests performed on a HDRB taken from the batch that was manufactured for the Solarino project (Markou et al. 2014). The tests were run at sev-

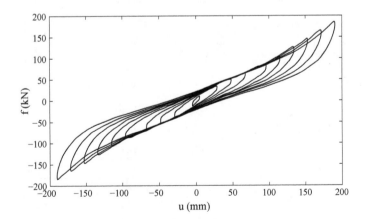

Fig. 7.3 Third cycle force-displacement response from harmonic tests on HDRB

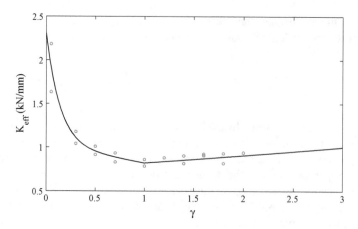

Fig. 7.4 Effective stiffness from harmonic tests on HDRB

eral amplitudes, up to a nominal shear strain $\gamma = 2$ ($T_r = 96\,\text{mm}$), at a frequency of 0.5 Hz, and a constant pressure of 6 MPa. A first series of tests was carried out with increasing displacement amplitudes, and then the same series was reversed. The results in terms of effective stiffness, K_{eff}, and equivalent viscous damping, ζ_{eq}, are shown in Figs. 7.4 and 7.5, respectively. In Fig. 7.4, the data points above the fitted curve refer to the first series of tests, i.e. the one where the displacement amplitude was increased up to the maximum. On the other hand, the data points below the curve are associated with the series of tests performed with decreasing displacement amplitudes. A slight decrease in effective stiffness, K_{eff}, is seen in the second series of tests. The opposite trend is observed in Fig. 7.5. In this case, the data points below the fitted curve refer to the first series of tests, while those above the curve are

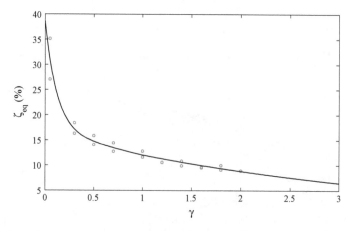

Fig. 7.5 Equivalent viscous damping from harmonic tests on HDRB

associated with the second. The equivalent viscous damping, ζ_{eq}, is seen to increase slightly in the second series of tests. In summary, the work performed on the isolator resulted in a slight decrease of effective stiffness and in a slight increase of equivalent viscous damping. For further details about testing of HDRBs, the reader is referred to ASCE/SEI 7 (2016), CEN (2009, 2005), Consiglio Superiore dei Lavori Pubblici (2008).

7.3   Existing Models for High Damping Rubber Bearings

The existing models used for the simulation of the behavior of HDRBs may be classified into two groups: (i) rate-independent, and (ii) rate-dependent. A review of the most important models for each category is presented in the following sections.

7.3.1   Rate-Independent Models

The simplest rate-independent model is represented by the bilinear system (Skinner et al. 1993), consisting of a linear spring connected in parallel with an elastic-perfectly plastic element (i.e. another linear spring connected in series with a plastic slider). An alternative mechanical model for the bilinear system is represented by a linear spring connected in series with the parallel system consisting of a linear spring and a plastic slider (Sivaselvan and Reinhorn 2006; Oliveto et al. 2014). In addition to the bilinear model, smooth hysteretic models exist, such as those proposed by Ozdemir (1976), and Wen (1976). In these models, the parameters that control the shape of the hysteresis loops are constant and determined at the beginning of the analysis. Fujita et al. (1990) improved upon the Ozdemir model by introducing a procedure to update parameters during the analysis. The main shortcoming of both the bilinear model and the smooth hysteretic models mentioned above is that they fail to adequately capture the nonlinear stiffening that elastomeric bearings exhibit in the large strain range.

Sanò and Di Pasquale (1995) modified a simple constitutive model for soils, the Martin-Davidenkov model (Martin and Bolton Seed 1978), to simulate the dynamic behavior of elastomeric devices. The model is characterized by 4 parameters, determined through experimental stress-strain hysteretic cycles at 0.1 Hz frequency and different values of maximum shear deformation between 7 and 50%. Comparison with the experimental curves shows that the proposed model fails to account for the observed energy dissipation at very low deformations.

Ahmadi et al. (1996) proposed a nonlinear hysteretic model for HDRBs, based on the observed stress-strain loop of a high damping natural rubber compound tested in shear at a frequency of 0.5 Hz and at a strain amplitude of 300%. The model can predict the stress-strain response reasonably well for amplitudes from 50 to 300%.

However, it underestimates the energy dissipated at small strain amplitudes, suggesting the presence of a rate-dependent dissipation mechanism.

Kikuchi and Aiken (1997) modified the model derived by Fujita et al. (1990) to account for stiffening of the bearings at high shear strain levels, and stiffness degradation between the first cycle and subsequent cycles at the same displacement amplitude. Hysteresis rules were developed for both steady-state motion and randomly-varying displacement conditions, typical of earthquake excitation. However, the model neglects the effects, on the response of the bearings, of strain-rate and variation of axial load.

Koo et al. (1996) proposed a modified bilinear model to consider the hardening behavior of elastomeric bearings at high strains. Koo et al. (1998) then extended such model to account for parametric variations of yield force and post-yield stiffness as functions of the maximum cyclic shear deformations.

Abe et al. (2004) extended the rate-independent elastoplastic model without degradation developed by Ozdemir (1976) by adding a displacement-dependent hardening rule to model the behavior at shear strains higher than 150%, and a non-linear elastic spring for strains up to 50%. The resulting one-dimensional model was then extended to consider biaxial shear deformation by means of a three-dimensional elastoplastic constitutive law of the Ozdemir model (Graesser and Cozzarelli 1989).

Grant et al. (2004) developed a strain-independent phenomenological model to describe the bidirectional shear force-deformation response of HDRBs. The restoring force vector is decomposed into a component parallel to the displacement vector and another parallel to the velocity vector. The former is a nonlinear elastic force defined by an odd, fifth order polynomial in the shear displacement, motivated by a generalized Mooney-Rivlin energy density function. The latter is a hysteretic response defined by an approach similar to bounding surface plasticity. Furthermore, two damage parameters are used to account for long-term scragging degradation and short term Mullins' effect (Mullins 1969).

The properties of elastomeric bearings are affected by variations of the imposed vertical load, due for instance to overturning forces during extreme ground shaking, and depend on the interaction between shear and axial forces. Strong variations generally occur for corner bearings in seismically isolated buildings and/or bearings experiencing large shear deformations. Under these conditions, elastomeric bearings exhibit stiffening behavior for low levels of compressive stress and buckling for high levels of compressive stress. Yamamoto et al. (2009) proposed an analytical model for elastomeric bearings that includes interaction between shear and axial forces, nonlinear hysteretic behavior at large shear strains, and dependence on vertical load. The model extends the one by Kikuchi and Aiken (1997) by including a discrete distribution of nonlinear axial springs at the top and bottom boundaries of the bearing. Its performance is illustrated by comparison with the results of cyclic experiments conducted on LRBs up to shear strains of 400% and different vertical load conditions from zero to 30 MPa. In some cases, the results are seen to differ from those given by other existing models.

Kikuchi et al. (2010) extended the above mentioned model by Yamamoto et al. (2009) to three dimensions, and applied it to predict the large shear deformation response of square elastomeric isolation bearings. The model, comprising multiple shear springs at mid-height and a series of axial springs at the top and bottom boundaries, was validated through cycling loading tests of LRBs under different vertical loads, and for different horizontal loading directions. An element (Kikuchi Bearing Element) based on this model is currently available in OpenSees (McKenna et al. 2000).

Yamamoto et al. (2012) carried out horizontal bidirectional loading tests on two real-sized HDRBs. The results show that the force in the primary direction of loading increases when there is displacement in the orthogonal direction. Moreover, bidirectional loading induces twist in the bearings, thus increasing local shear strains. Based on these observations, the authors proposed a rate-independent nonlinear mathematical model, which simulates reasonably well the test results under both bidirectional and unidirectional loading. The model is similar to the one proposed by Abe et al. (2004). Here, however, the restoring force is decomposed into two separate non-orthogonal components, a nonlinear elastic spring directed to the origin and a variable friction element, approximately parallel to the direction of motion, accounting for energy dissipation. The model requires a limited number of parameters, as compared to the 13 required by the model proposed by Abe et al. (2004), but relies on the numerical integration of a first order vector differential equation.

The Bridgestone Corporation created the next-generation high damping rubber bearings. Compared to traditional HDRBs, they are influenced less by loading history, and exhibit a more stable hysteretic behavior under cyclic loading at large shear deformations (Bridgestone Catalog 2013). The manufacturer provides a set of polynomial equations relating effective shear modulus, equivalent damping ratio, and ratio of characteristic strength to maximum shear force in a cycle to shear strain, γ, in the range from 10 to 270%. These can then be used to evaluate (i) effective stiffness and equivalent damping ratio, to be used in an equivalent linear model, (ii) initial stiffness, post-yield stiffness and characteristic strength, to be used in a bilinear idealization. Analytical expressions are also provided for the dependency of effective stiffness, and equivalent damping ratio, on temperature varying between -10 and $40\,°C$.

Gjorgjiev and Garevski (2013) developed a simple 1D analytical model for the simulation of the nonlinear force-displacement relationship of low and high damping rubber bearings. The restoring force is expressed as a polynomial function of bearing deformation. The behavior of the bearing is presented through the polynomial coefficients plus an additional eight parameters, obtained from bi-axial tests. The model is capable of capturing the hardening of rubber at large deformations, but does not account for the Mullins' effect, strain-rate, and dependence on the axial load. The authors suggest the use of property modification factors to capture the effects of ageing, temperature and scragging. The analytical model was verified through biaxial tests on square and circular bearings produced from different rubber compounds. The results show its inability to accurately predict the behavior of the bearings at low deformations, due to the fact that it was calibrated at large strains.

Kato et al. (2014) proposed a new analytical model for the evaluation of elastoplastic and creep-like behavior of HDRBs under bidirectional seismic and strong wind excitation. The authors use the constitutive law proposed by Mori et al. (2010, 2011). One of its important features is that the creep-like behavior is represented by an elastoplastic expression rather than a viscoelastic one. In a viscoelastic constitutive law, stress relaxation is a function of time. On the other hand, in the constitutive law developed by Mori et al. (2010, 2011), stress relaxation is a function of deformation history, which is expressed by the increment of the cumulative equivalent strain. The first two terms of the Yeoh model (Yeoh 1993), and the Neo-Hookean model (Ogden 1998) are used as strain energy density functions related to elasticity and plasticity, respectively. The effects of axial force on the horizontal stiffness of the bearing are not considered by the model.

Markou et al. (2016) recently presented a trilinear hysteretic model defined by 5 parameters. The model accounts for hardening at high strain amplitudes, and consists of a linear elastic spring, connected in series with a parallel system composed of a plastic slider and a trilinear elastic spring. The model was calibrated through a series of free vibration tests from the Solarino project (Oliveto et al. 2004) performed at shear strains up to 140%.

7.3.2   Rate-Dependent Models

Nagarajaiah et al. (1989) suggested the use of a modified viscoplastic model, originally proposed by Bouc (1967), and subsequently extended by Wen (1976). According to this model, the restoring force is given by a linear elastic component and a hysteretic elastoplastic component. The latter depends on a hysteretic dimensionless parameter governed by a nonlinear first order differential equation. Coupled differential equations were used to model inelastic bi-axial behavior, considering interaction between forces in two orthogonal horizontal directions. Shear stiffness degradation observed in HDRBs is accounted for by varying the post yielding to elastic stiffness ratio. This bi-axial hysteretic model is implemented in Program 3D-BASIS (Nagarajaiah et al. 1989).

Tsopelas et al. (1994) extended the model developed by Nagarajaiah et al. (1989) to account for hardening of rubber at large shear strains. To this end, a nonlinear elastic spring was included in the model. The behavior of the spring is linear in the low and high shear strain ranges, and quadratic in between. The 2D version of the model is available in Program 3D-BASIS.

Pan and Yang (1996) proposed a 1D mathematical model for HDRBs, to be used in the nonlinear seismic analysis of base-isolated multistory buildings. The shear force in the bearings is represented as a nonlinear function of shear deformation and velocity. The 11 parameters needed to define the model can be determined from cyclic tests of full-scale bearings.

Hwang and Ku (1997) proposed two analysis models for HDRBs based on the results of shaking table tests of a seismically isolated bridge deck. The bearings were strained up to a maximum shear of 100%. In the first model, effective stiffness and equivalent damping ratio are determined using a system identification method. In the second, a fractional derivative Kelvin model is used. The shear modulus of the bearings is predetermined according to the AASHTO (1991) specifications, and then used to determine the parameters involved in the fractional derivative Kelvin model. The two proposed models appear to predict the seismic response of the test structure more accurately than the linear equivalent models, implemented as recommended by AASHTO (1991) specifications and the JPWRI (1992) manual. In practical applications, the maximum shear strain of the bearings under a design earthquake is not known prior to analysis. Therefore, an iteration procedure is illustrated, that can be implemented, when using the fractional derivative Kelvin model, to determine the maximum shear strain needed to evaluate the effective shear modulus. Hwang and Hsu (2001) introduced temperature in the model, and carried out a series of tests with temperature varying from 0 to 28 °C.

Hwang and Wang (1998) carried out a similar experimental campaign, extended to shear strains up to about 200%, and proposed a fractional derivative Maxwell model. The parameters of the model were obtained through identification of harmonic tests based on the best fit of the dynamic amplification factor, or the phase angle. The comparison between measured and predicted responses indicates that the parameters should be determined by matching the phase angle rather than the dynamic amplification factor.

Naeim and Kelly (1999) reported that whereas the equivalent viscous damping ratio typically decreases with increasing strain, energy dissipation per cycle does not. Experiments show that the energy dissipated in one cycle actually increases and is proportional to the shear strain raised to an exponent of approximately 1.5. Based on this observation, the authors suggested an energy-based model for elastomeric isolation systems consisting of a linear elastic spring, a pure hysteretic element (energy dissipation proportional to displacement D), and a pure viscous element (energy dissipation proportional to D^2). The properties of each of these elements are to be determined so that the energy dissipated by the combined system is proportional to $D^{1.5}$ over a given strain range. In order to account for rubber stiffening at high strain amplitudes, the authors propose a gap/stiffening element with separate loading and unloading curves, guaranteeing energy dissipation in the stiffening portion of the loops. The authors state that most high-damping natural rubbers do not exhibit significant rate-dependence within the range of frequencies anticipated in seismic applications, and show that relatively little change in energy dissipation may be expected in the bearings.

Hwang et al. (2002) modified and extended the mathematical model proposed by Pan and Yang (1996) to include the capability of describing the effects of scragging, Mullins, frequency and temperature. The model is defined by 10 parameters, to be determined from cyclic loading tests. A sensitivity analysis is presented to show how the behavior is affected by changes of the parameters. A set of cyclic material tests and shake table tests in the form of harmonic tests and earthquake simulations are carried out to validate the model. Further studies are needed to establish a generalized relationship between the parameters of the proposed model and the different factors characterizing the behavior of HDRBs.

Jankowski (2003) presented an 11-parameter model along the lines of those proposed by Pan and Yang (1996) and Hwang et al. (2002). The model was used to simulate three sets of cyclic tests on three different rubber bearings. An optimal set of parameters was identified for each set of tests and the identification error provided.

Tsai et al. (2003) proposed an analytical model for HDRBs based on the models originally developed by Bouc (1967) and Wen (1976), and subsequently extended to two directions by Park et al. (1986). Wen's model is presented in incremental form and modified to adequately simulate the strain hardening of rubber at high shear strains, and rate-dependent effects. The restoring force in the bearings is given as the summation of a displacement dependent shear force and a velocity dependent shear force. The model was validated through a series of cyclic tests at different strain amplitudes, frequencies and applied vertical loads.

Dall' Asta and Ragni (2006) proposed a nonlinear viscoelastic damage model to describe the behavior of high damping rubber under cyclic loading. The model was developed to adequately capture the observed transient (Mullins' softening effect) and stable responses of rubber in the range of strain rates and amplitudes of interest for seismic applications. Internal variables and their evolution laws are introduced to describe the inelastic phenomena in the transient and stable phases of the response. Analytical results show a reasonably good agreement with experimental data.

Bhuiyan et al. (2009) developed a rate-dependent elasto-viscoplastic model suitable for seismic analysis of a bridge isolated with HDRBs. The model consists of a Maxwell model connected in parallel with a nonlinear elastic spring and an elasto-plastic model (spring-slider). The constitutive relations for each component of the rheological model are identified through a set of experiments at room temperature, including a cyclic shear test, a multi-step relaxation test, and a simple relaxation test. In particular, a nonlinear viscosity law is deduced for the dashpot of the Maxwell model, which is capable of reproducing the rate-dependent behavior of HDRBs.

Kumar et al. (2014) addressed many of the issues involved in modeling the response of low damping rubber bearings, and lead rubber bearings, under loadings associated with extreme and beyond design basis earthquakes. These include coupled bidirectional motion in horizontal directions, coupling of vertical and horizontal motion, cavitation and post-cavitation behavior in tension, strength degradation in cyclic tensile loading due to cavitation, variation in critical buckling load capacity due to lateral displacement, and for LRBs, strength degradation in cyclic shear loading due to heating of the lead core. The authors proposed an integrated numerical model capable of addressing these issues. The model consists of a 2-node, 12 degrees-of-freedom, discrete element. The two nodes are connected by six springs, representing the mechanical behavior along the 6 principal directions of a bearing. The two shear springs are coupled using the 2-dimensional Bouc-Wen model (Bouc 1967; Wen 1976; Nagarajaiah et al. 1989). All the other springs are uncoupled. The coupling of horizontal and vertical motions is considered indirectly by using expressions for mechanical properties in one direction depending on response parameters in the other direction. The model was verified and validated following ASME best practices (ASME V&V 10-2006 2006) and implemented in OpenSees (Kumar et al. 2014).

Nguyen et al. (2015) carried out an experimental program to investigate the behavior of HDRBs at very low temperatures. Based on the results of the experiments, they developed an elasto-viscoplastic rheological model similar to the one proposed by Bhuiyan et al. (2009). In this case, an expanded Maxwell model is used to describe the rate-dependent behavior of the bearings, that is the linear elastic spring of the Maxwell model is replaced by a bilinear system. The model was reasonably successful in reproducing the typical rate-dependent cyclic behavior of HDRBs under various testing conditions. However, comparison between experimental data and numerical results points out some obvious differences. These may be attributed to phenomena like healing of the Mullins' softening effect, self-heating, and their respective temperature history dependences, effects that are not considered in the rheological model.

Markou and Manolis (2016) suggested a 4-parameter fractional derivative Zener model for the simulation of the behavior of HDRBs under shear deformation. The mechanical model is given by a linear spring, connected in series with a fractional derivative viscous element in parallel with a linear spring. The authors calibrated the model at a maximum shear strain of 140%, and used it to simulate the free vibration tests carried out as part of the Solarino project (Oliveto et al. 2004). The results were compared to those obtained with the Zener model, the bilinear model, and the trilinear model (Markou et al. 2016). Whereas the proposed model appears to be qualitatively and quantitatively better than the Zener model, its performance is well below that of the bilinear and trilinear hysteretic models.

7.4 Discussion

The literature review of the previous section gives a hint of the complexity of the mechanical behavior of HDRBs. A single comprehensive model capable of tackling all aspects of this complex behavior may be unfeasible, too sophisticated, or in many ways impractical. Based on the particular application, the use of simpler models, dealing with specific aspects of the behavior, might be preferable.

The models that are currently available in the literature can be classified as rate-dependent and rate-independent. Shear cyclic tests with strain amplitudes of the order of 100% indicate that quite different force-deformation responses are obtained if bearings are tested under dynamic or quasi-static conditions. Oliveto et al. (2013) show that, in a dynamic test at a frequency of 0.5 Hz, increases of 22% in post-yield stiffness, and 37% in force at zero displacement, are obtained with respect to the quasi-static test on the same bearing. On one hand, these results suggest substantial rate-dependency. On the other, cyclic tests at multiple loading frequencies in the range of interest for seismic applications show negligible differences in terms of force-deformation response, thus justifying the large number of rate-independent models available in the literature.

Creep and stress-relaxation are two typical phenomena exhibited by HDRBs. Upon applying a load, it may take some time for the system to reach its final equilibrium configuration. On the other hand, when a strained configuration is imposed, it will take time to reach the final state of stress. These phenomena can only be described by rate-dependent models, thus explaining the abundance of such models in the literature.

HDRBs, like any other kind of bearing, carry vertical loads due to gravity, and additional vertical loads induced by wind and earthquakes. The variation of vertical load due to earthquakes can be particularly significant in tall buildings, and in some cases can result in tension loads for some bearings. The compression seismic overload may cause buckling of the bearing, while tension may determine cavitation in the rubber layers. Seismic regulations tend to prevent tension loads in bearings, and to ensure that buckling loads are sufficiently large as to not affect the load-carrying capacity in compression, even under large shear deformation. However, a model for the description of the mechanical behavior of HDRBs should be able to consider the variation of the axial load due to earthquakes and provide a reliable response under the considered earthquake. Occasionally, in order to prevent tension, designers have allowed for some limited uplift of laminated natural rubber bearings (Kikuchi et al. 2014). Axial forces in bearings are also affected by vertical ground accelerations, requiring the development of models that account for the simultaneous action of horizontal and vertical ground motions.

Recent experimental research has shown that bi-directional horizontal ground motion can result in stress states that are more severe than those that are induced by one-directional motions with the same displacement amplitude. The reason for this behavior is attributable to torsion of the bearings caused by the bi-directional horizontal ground motion, explaining the need to extend the available 1D mechanical models to two directions, in order to account for the twisting effect.

One of the issues that has been thoroughly addressed in the literature is the stiffening of rubber at large shear deformations, a phenomenon which cannot be described by the simple and popular bi-linear model. Several models have been proposed, involving different characterization of both the non-linear elastic and energy dissipation components of the response.

An additional aspect, discussed to some extent, is the bearings' stiffness and strength degradation associated with loading history. In cyclic loading under displacement control, the stiffness and strength of an untested bearing are considerably larger in the first cycle than in subsequent cycles. After the third cycle, the hysteresis loops are almost coincident and may be considered so for practical applications. Seismic regulations often recommend to refer to the third cycle, when establishing basic properties of bearings such as effective stiffness and equivalent damping ratio. In recent literature (Clark et al. 1997), the stiffness and strength degradation occurring in the first cycle of loading at a given shear amplitude is denoted as scragging, while the slow degradation in following cycles is denoted as Mullins' effect. However, the latter denomination has been widely used in the past to describe the overall phenomenon. It is obvious that, in the first cycle of response when an earthquake occurs, models that are calibrated with reference to the third cycle of cyclic laboratory tests will underestimate forces, resulting in the overstressing of the superstructure. Furthermore, under long-duration earthquakes, bearings can undergo many cycles at large strains, resulting in stiffness softening and strength degradation typical of the Mullins' effect. Repeated cycles may also cause heating of the bearings, especially in LRBs, but to a minor extent also in HDRBs. A model that does not account for the Mullins' effect may underestimate the displacement of the bearings under long-duration earthquakes and overestimate the response of the superstructure. Whereas the Mullins' effect recedes in the short term, there is evidence that scragging also disappears in the long run and that rubber eventually recovers its original and unscragged state. Even though bearings generally undergo scragging during the acceptance tests, they may have recovered their unscragged properties several years later when an earthquake occurs. Therefore, it is important that models for HDRBs be capable of describing scragging and the Mullins' effect adequately.

Another issue that ought to be considered is the dependence of HDRBs' properties on temperature. During its lifetime, the temperature of an isolator may undergo large excursions, and the properties corresponding to the upper bound of the temperature range may be considerably different from the properties corresponding to the lower bound. Seismic codes usually require that the designer assess the design in the two extreme conditions, and the models discussed in the previous section generally require that a different set of parameters be defined for each state. It may be worth pointing out that each one of these sets of parameters needs to be identified from tests performed at the corresponding temperature. Isolators to be used in cold climates, such as in sub-polar areas, are usually tested in environmental test chambers.

Most of the models available in the literature are mathematically based, and are not supported by a clear-cut and rigorous mechanical formulation, enabling their use under general loading. In fact, a large number of models are derived by fitting cyclic experimental data at a given strain amplitude, and require a different set of parameters when the amplitude of deformation changes. In most cases, the extension to general loading cases is either not clear or extremely cumbersome.

The ASCE/SEI 7 (2016) proposes the use of lower and upper bounds for the nominal properties of seismic isolation bearings. The use of upper bounds leads to lower values of relative displacement of the isolators and to larger values of the force transmitted to the superstructure. On the other hand, the use of lower bounds leads to larger relative displacements for the isolators and lower forces for the superstructure. The designer will alternatively use upper or lower bounds based on whether the safety of the superstructure or that of the isolator is being assessed. A general procedure for establishing the upper and lower bound properties of seismic isolators through the use of modification factors is illustrated by McVitty and Constantinou (2015).

In the following section, a new model is presented, based on the combination of a number of well-established physical models, and allowing for the cyclic behavior of HDRBs to be simulated at low, intermediate and large strains. Furthermore, its extension to earthquake excitation is simple and straightforward, since it relies on well-known physical models extensively tested to that purpose. Presently, the model is limited to 1D motion. Extension to 2D motions, and the inclusion of other effects, such as axial force variation, are subject of ongoing and future work.

7.5 An Alternative Model for High Damping Rubber Bearings

As shown in Fig. 7.6, the new 10-parameter rheological model suggested for the simulation of the shear behavior of HDRBs consists of four elements: (a) a nonlinear elastic spring (element 1), (b) two elastoplastic elements (elements 2 and 3) and (c) a hysteretic damping element (element 4).

The restoring force in the nonlinear spring is a linear function of displacement, for amplitudes below u_a, and a quadratic function of displacement for amplitudes larger than u_a. In mathematical form it may be expressed as

$$f_{e1} = \begin{cases} k\, u; & \text{if } |u| \leq u_a \\ f_0\, sgn(u) + k_1\, u + k_2\, u^2\, sgn(u); & \text{if } |u| \geq u_a \end{cases} \tag{7.6}$$

where sgn is the sign function. By expressing k_1 and k_2 as

$$k_1 = k - \frac{2f_0}{u_a}; \quad k_2 = \frac{f_0}{u_a^2} \tag{7.7}$$

it follows that only three parameters are needed to define the nonlinear elastic spring element, i.e. k, u_a, and f_0. Each of the two elastoplastic elements is a two-parameter system defined in terms of the elastic stiffness, k_{0i}, and the characteristic strength, Q_i ($i = 2, 3$). The yield displacements, u_{yi} ($i = 2, 3$), can be defined in terms of the previous two parameters as follows:

$$u_{yi} = \frac{Q_i}{k_{0i}}; \quad i = 2, 3 \tag{7.8}$$

The forces in the two elastoplastic elements are then described by

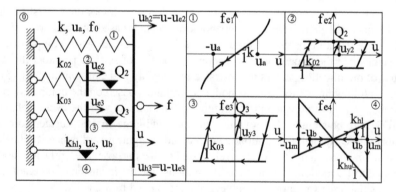

Fig. 7.6 Proposed rheological model for HDRBs

Table 7.1 Identified model parameters

$u_{y2}(m)$	Q_2 (kN)	u_{y3} (m)	Q_3 (kN)	k (kN/m)	u_a (m)	f_0 (kN)	k_{hl} (kN/m)	u_b (m)	u_c (m)	e^2 (%)
0.00031	4.378	0.01702	10.728	696.079	0.102	47.081	70.818	0.134	0.0267	5.32

$$f_{ei} = \begin{cases} Q_i \, sgn(\dot{u}_{hi}); & \text{if } \dot{u}_{hi} \neq 0 \\ k_{0i} \, u_{ei}; & \text{if } \dot{u}_{hi} = 0 \end{cases} \quad i = 2,3 \tag{7.9}$$

The behavior of the hysteretic damper is described by three parameters, namely k_{hl}, u_c and u_b, and the constitutive equation of the element is given by

$$f_{e4} = \frac{1}{2} k_{hl} u \, (1 + sgn(u\dot{u})) + H(u_m - u_b)\frac{1}{2}k_{hu} u \, (1 - sgn(u\dot{u})) \tag{7.10}$$

where H is the Heaviside function, u_m is the absolute value of the displacement at the end of the last loading phase, and

$$k_{hu} = -k_{hl} \left(\frac{u_m - u_b}{u_c} \right) \tag{7.11}$$

Loading phases are characterized by $sgn(u\dot{u}) > 0$, while unloading phases are identified by $sgn(u\dot{u}) < 0$.

Finally, the overall force in the system is given by

$$f = f_{e1} + f_{e2} + f_{e3} + f_{e4} \tag{7.12}$$

The model was calibrated using a set of cyclic tests at 10 different strain amplitudes ($0.05 \leq \gamma \leq 2$), and at a frequency of 0.5 Hz, performed on a HDRB from the Solarino project (Markou et al. 2014). The identification process was carried out using an evolutionary algorithm, namely the covariance matrix adaptation-evolution strategy (CMA-ES) (Hansen 2011), and the identified parameters are given in Table 7.1. The identification error, e^2, shown in Table 7.1, is defined as follows:

$$e^2 = \frac{\langle F_0 - \widetilde{F}, F_0 - \widetilde{F} \rangle}{\langle F_0, F_0 \rangle} \tag{7.13}$$

where F_0, and $\widetilde{F}$, are the measured and computed force vectors, and

$$\langle A, B \rangle = \sum_{i=1}^{n} A_i B_i \tag{7.14}$$

is the standard inner product.

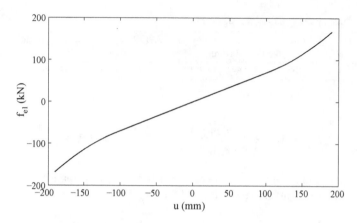

Fig. 7.7 Force-displacement response of nonlinear spring

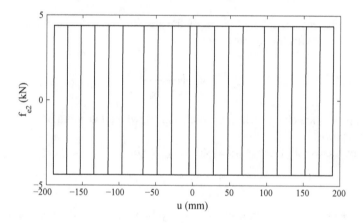

Fig. 7.8 Force-displacement response of first elastoplastic element (element 2)

The force-displacement plots for the four components of the model are shown separately in Figs. 7.7, 7.8, 7.9 and 7.10, while the force-displacement response of the overall model is given in Fig. 7.11. A comparison between simulated and experimental force-displacement curves is provided in Fig. 7.12.

It may be instructive to illustrate how the four in-parallel elements of the model behave at different displacement amplitudes and in different phases of motion (loading and unloading), and how they finally constitute the overall behavior of the proposed model. The system has 4 characteristic displacements: $u_{y2} < u_{y3} < u_a < u_b$. For displacements smaller than the yield displacement of element 2 ($|u| < u_{y2}$), which is infinitesimal (see Table 7.1), the system responds with stiffness equal to $k + k_{02} + k_{03} + k_{hl}$ in a loading phase, and $k + k_{02} + k_{03}$ during unloading. Following yielding of element 2, and up to yielding of element 3 ($u_{y2} < |u| < u_{y3}$), the stiffness of the system will be equal to $k + k_{03} + k_{hl}$ in the loading phases, and $k + k_{03}$ when

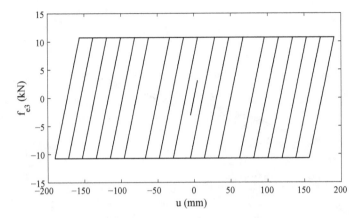

Fig. 7.9 Force-displacement response of second elastoplastic element (element 3)

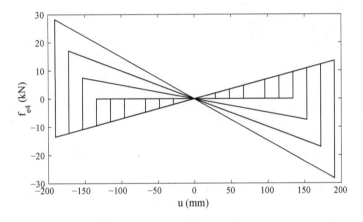

Fig. 7.10 Force-displacement response of hysteretic damper

unloading. It may be worth mentioning that the elastic stiffness k_{03}, in addition to k and k_{hl}, takes care of the stiff behavior exhibited by HDRBs at low strain amplitudes. After element 3 has yielded, and up to u_a ($u_{y3} < |u| < u_a$), the stiffness of the system is equal to $k + k_{hl}$ in loading, and simply k during unloading. The characteristic displacement, u_a, denotes initiation of stiffness hardening, typically exhibited by HDRBs at high shear deformation. For the bearing considered ($T_r = 96$ mm), u_a corresponds approximately to a strain amplitude $\gamma = 1$. Thereafter, as shown in Fig. 7.4, the effective stiffness of the bearing, K_{eff}, starts to increase. The last characteristic displacement, u_b, introduces stiffness k_{hu} in the unloading phase of the hysteretic damper (element 4). Note that k_{hu} is zero at displacement amplitudes smaller than u_b. Governed by Eq. 7.11, k_{hu} increases in magnitude with increasing values of u_m. In summary, the nonlinear spring (element 1) is used to account for the observed stiffness hardening at high strain amplitudes ($|u| > u_a$), elastoplastic element 3, and

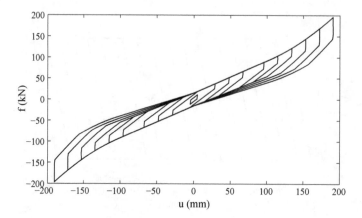

Fig. 7.11 Global force-displacement response

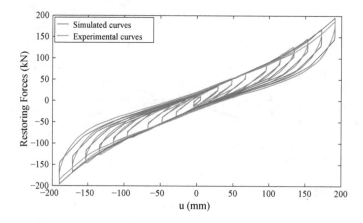

Fig. 7.12 Measured and simulated force-displacement response

in particular its elastic stiffness, k_{03}, describes the high stiffness at lower strain ampli-
tudes ($u_{y2} < |u| < u_{y3}$), where energy dissipation is still accounted for by elastoplas-
tic element 2, and the hysteretic damper (element 4) allows for different slopes of the
force-displacement response in loading and unloading phases.

7.6 Conclusions

After a brief introduction on the chemical composition, manufacturing, and mechanical properties of rubber bearings in general, the tests required on these devices by current seismic codes, and the main mechanical properties that are generally extracted
from such tests, have been described. An overview has been given of the most impor-

tant models available in the literature to describe the highly complex response of HDRBs to combined shear deformation and compression. The issues and uncertainties involved in the simulation of such behavior, as well as the shortcomings of existing models, have been pointed out.

In an effort to overcome these problems, recent codes suggest the use of simple models with upper and lower bound design properties. To this end, a new simple model has been proposed in this work, for the simulation of the shear behavior under compression of HDRBs. The model consists of four elements connected in parallel. Three of them are very simple and well-known elements, i.e. a nonlinear elastic spring, used to describe stiffness hardening and/or softening, and two elastoplastic elements accounting for energy dissipation. The fourth and last element is a hysteretic damper used to describe the different behavior of HDRBs during loading and unloading phases of dynamic response. The model simulates the shear behavior of HDRBs under a wide range of strain amplitudes, up to $\gamma = 2$, quite successfully, providing small identification error between the recorded and simulated force-displacement curves.

Acknowledgements This work was carried out with the financial support from ReLUIS (Italian National Network of University Earthquake Engineering Laboratories), Project D.P.C.-ReLUIS 2014–2016, WP1.

References

AASHTO (1991) Guide specifications for seismic isolation design. American Association of State Highway and Transportation Officials (AASHTO), Washington, DC

Abe M, Yoshida J, Fujino Y (2004) Multiaxial behaviors of laminated rubber bearings and their modeling. II: Modeling. J Struct Eng ASCE 130:1133–1144

Ahmadi HR, Fuller KNG, Muhr AH (1996) Predicting response of non-linear high damping rubber isolation systems. In: Proceedings of 11th world conference on earthquake engineering, Acapulco, Mexico, Paper no 1836

ASCE/SEI 7 (2016) Minimum design loads for buildings and other structures, Chapter 17. American Society of Civil Engineers (ASCE), New York

ASME V&V 10-2006 (2006) Guide for verification and validation in computational solid mechanics. American Society of Mechanical Engineers (ASME), New York

Bhuiyan AR, Okui Y, Mitamura H, Imai T (2009) A rheology model of high damping rubber bearings for seismic analysis: identification of nonlinear viscosity. Int J Solids Struct 46:1778–1792

Bouc R (1967) Forced vibration of mechanical systems with hysteresis. In: Proceedings of 4th conference on nonlinear oscillation, Prague, Czechoslovakia

Bozorgnia Y, Bertero VV (2004) Earthquake engineering, from engineering seismology to performance-based design. CRC Press, London

Bridgestone Catalog (2013) Seismic isolation product line-up, Tokyo

CEN (Comitè Europèen de Normalisation) (2005) EN 1337–3: Structural bearings—Part 3: Elastomeric bearings. European Committee for Standardization, Brussels, Belgium

CEN (Comitè Europèen de Normalisation) (2009) EN 15129: Anti-seismic devices. European Committee for Standardization, Brussels, Belgium

Ciesielski A (1999) An introduction to rubber technology. Rapra Technology Limited, Southampton

Clark PW, Aiken ID, Kelly JM (1997) Experimental studies of the ultimate behavior of seismi-cally isolated structures. Report no. UCB/EERC-97/18, Earthquake Engineering Research Center, University of California, Berkeley, California

Consiglio Superiore dei Lavori Pubblici (2008) D.M. 14 gennaio 2008, Norme tecniche per le costruzioni, Italy

Constantinou MC, Whittaker AS, Kalpakidis Y, Fenz DM, Warn GP (2007) Performance of seismic isolation hardware under service and seismic loading. In: Technical report MCEER-07-0012, State University of New York at Buffalo

Dall' Asta A, Ragni L (2006) Experimental tests and analytical model of high damping rubber dissipating devices. Eng Struct 28:1874–1884

FEMA P-751 (2009) NEHRP recommended seismic provisions: design examples, Chapter 12. Kircher, CA

Fujita T, Suzuki S, Fujita S (1990) High damping rubber bearings for seismic isolation of buildings (1st report, hysteretic restoring force characteristics and analytical models). Trans Jpn Soc Mech Eng C56:658–666 (in Japanese)

Gent A (2012) Engineering with rubber. Hanser Publications, Munich

Gjorgjiev I, Garevski M (2013) A polynomial analytical model of rubber bearings based on series of tests. Eng Struct 56:600–609

Graesser EJ, Cozzarelli FA (1989) Multidimensional models of hysteretic material behavior for vibration analysis of shape memory energy absorbing devices. Technical report NCEER-89-0018, State University of New York at Buffalo

Grant DN, Fenves GL, Whittacker AS (2004) Bidirectional modeling of high damping rubber bearing. J Earthq Eng 8(1):161–185

Hansen N (2011) The CMA evolution strategy: a tutorial. https://www.lri.fr/~hansen/cmatutorial. pdf

Hwang JS, Wu JD, Pan TC, Yang G (2002) A mathematical hysteretic model for elastomeric isolation bearings. Earthq Eng Struct Dyn 31:771–789

Hwang JS, Hsu TY (2001) A fractional derivative model to include effect of ambient temperature on HDR bearings. Eng Struct 23:484–490

Hwang JS, Ku SW (1997) Analytical modeling of high damping rubber bearings. J Struct Eng ASCE 123(8):1029–1036

Hwang JS, Wang JC (1998) Seismic response prediction of HDR bearings using fractional derivative Maxwell model. Eng Struct 30(9):849–856

Jankowski R (2003) Nonlinear rate dependent model of high damping rubber bearing. Bull Earthq Eng 1:397–403

Japanese Public Works Research Institute (JPWRI) (1992) Manual for Menshin design of highway bridges. Ministry of Construction, Tsukuba Science City, Japan (in Japanese)

Kato H, Mori T, Murota N, Kikuchi M (2014) Analytical model for elastoplastic and creep-like behavior of high-damping rubber bearings. J Struct Eng ASCE. doi:10.1061/(ASCE)ST.1943-541X.0001181

Kelly JM, Konstantinidis DA (2011) Mechanics of rubber bearings for seismic and vibration isolation. Wiley, New York

Kikuchi M, Nakamura T, Aiken ID (2010) Three-dimensional analysis for square seismic isolation bearings under large shear deformations and high axial loads. Earthq Eng Struct Dyn 39:1513–1531

Kikuchi T, Takcuchi T, Fujimori S, Wada A (2014) Design of seismic isolated tall building with high aspect-ratio. Int J High-Rise Build 3(1):1–8

Kikuchi M, Aiken ID (1997) An analytical hysteresis model for the elastomeric seismic isolation bearings. Earthq Eng Struct Dyn 26:215–231

Koo GH, Lee JH, Kim JB, Lee HY, Yoo B (1996) Reduction of seismic responses by using the modified hysteretic bi-linear model of the seismic isolator. Trans Korea Soc Mech Eng 20(1):127–134

Koo GH, Lee JH, Yoo B (1998) Seismic response analyses of seismically isolated structures using laminated rubber bearings. J Korean Nucl Soc 30(5):387–395

Kumar M, Whittaker AS, Constantinou MC (2014) An advanced numerical model of elastomeric seismic isolation bearings. Earthq Eng Struct Dyn 43:1955–1974

Markou AA, Oliveto G, Athanasiou A (2016) Response simulation of hybrid base isolation systems under earthquake excitation. Soil Dyn Earthq Eng. doi:10.1016/j.soildyn.2016.02.003

Markou AA, Oliveto G, Mossucca A, Ponzo FC (2014) Laboratory experimental tests on elastomeric bearing from the Solarino project, Progetto di Ricerca DPC - RELUIS, Linea di Ricerca 6: Isolamento e Dissipazione. Ponzo FC and Serino G, Coordinatori

Markou AA, Manolis GD (2016) A fractional derivative Zener model for the numerical simulation of base isolated structures. Bull Earthq Eng 14(1):283–295

Martin PP, Bolton Seed H (1978) MASH, A computer program for the non-linear analysis of vertically propagating shear waves in horizontally layered deposits. Report UCB/EERC 78/23. University of California, Berkeley, CA

McKenna F, Fenves GL, Scott MH, Jeremic B (2000) Open system for earthquake engineering simulation (OpenSees). Pacific Earthquake Engineering Research Center, University of California, Berkeley, CA

McVitty WJ, Constantinou MC (2015) Property modification factors for seismic isolators: design guideline for buildings. Technical report MCEER-15-0005, State University of New York at Buffalo

Michalovic M, Brust GJ (2000) The story of rubber, a self-guided polymer expedition. http://pslc.ws/macrog/exp/rubber/menu.htm

Mori T, Kato H, Murota N (2010) FEM analysis of high damping laminated rubber bearings using anelastic-plastic constitutive law of the deformation history integral type. J Struct Constr Eng 75(658):2171–2178 (in Japanese)

Mori T, Kato H, Kikuchi T, Murota N (2011) Elastic-plastic constitutive law of rubber for laminated rubber bearings. In: Proceedings of 12th world conference on seismic isolation, energy dissipation and active vibration control of structure, ASSIS, Sochi, Russia

Mullins L (1969) Softening of rubber by deformation. Rubber Chem Technol 42(1):339–362

Naeim F, Kelly JM (1999) Design of seismic isolated structures: from theory to practice. Wiley, New York

Nagarajaiah S, Reinhorn AM, Constantinou MC (1989) Nonlinear dynamic analysis of three-dimensional base isolated structures (3D-Basis). Technical report NCEER-89-0009, State University of New York at Buffalo

Nguyen DA, Dang J, Okui Y, Amin AFMS, Okada S, Imai T (2015) An improved rheology model for the description of the rate-dependent cyclic behavior of high damping rubber bearings. Soil Dyn Earthq Eng 77:416–431

Ogden RW (1998) Nonlinear elastic deformations. Elasticity. Dover Publications, Mineola, NY

Oliveto G, Oliveto ND, Athanasiou A (2014) Constrained optimization for 1-D dynamic and earthquake response analysis of hybrid base-isolation systems. Soil Dyn Earthq Eng 67:44–53

Oliveto G, Athanasiou A, Granata M (2013) Blind simulation of full scale free vibration tests on a three story base isolated building. In: Proceedings of 10th international conference on urban earthquake engineering, Tokyo

Oliveto G, Granata M, Buda G, Sciacca P (2004) Preliminary results from full-scale free vibration test on a four story reinforced concrete building after seismic rehabilitation by base isolation. In: Proceedings of JSSI 10th anniversary symposium on performance of response controlled buildings, Yokohama

Ozdemir H (1976) Nonlinear transient dynamic analysis of yielding structures. PhD dissertation, University of California, Berkeley, California

Pan TC, Yang G (1996) Nonlinear analysis of base isolated MDOF structures. In: Proceedings of 11th world conference on earthquake engineering, Acapulco, Mexico, Paper No. 1534

Park YJ, Wen YK, Ang AH-S (1986) Random vibration of hysteretic systems under bi-directional ground motions. Earthq Eng Struct Dynmics 14:543–557

Sanò T, Di Pasquale G (1995) A constitutive model for high damping rubber bearings. J Press Vessel Technol ASCE 107:53–58

Shenton HW (1996) NISTIR 5800, guidelines for pre-qualification, prototype and quality control testing of seismic isolation systems. National Institute of Science and Technology (NIST), Gaithersburg, Maryland

Sivaselvan MV, Reinhorn AM (2006) Lagrangian approach to structural collapse simulation. J Eng Mech ASCE 132(8):795–805

Skinner RI, Robinson WH, McVerry GH (1993) An introduction to seismic isolation. Wiley, Chistester

Treolar LRG (1975) The physics of rubber elasticity. Clarendon Press, Oxford

Tsai CS, Chiang TC, Chen BJ, Lin SB (2003) An advanced analytical model for high damping rubber bearings. Earthq Eng Struct Dyn 32:1373–1387

Tsopelas PC, Constantinou MC, Reinhorn AM (1994) 3D-BASIS-ME: computer program for nonlinear dynamic analysis of seismically isolated single and multiple structures and liquid storage tanks. Technical report NCEER-94-0010, State University of New York at Buffalo

Wen YK (1976) Method for random vibration of hysteretic systems. J Eng Mech ASCE 102:249–263

Yamamoto S, Kikuchi M, Ueda M, Aiken ID (2009) A mechanical model for elastomeric seismic isolation bearings including the influence of axial force. Earthq Eng Struct Dyn 38:157–180

Yamamoto M, Minewaki S, Yoneda H, Higashino M (2012) Nonlinear behavior of high-damping rubber bearings under horizontal bidirectional loading: full-scale tests and analytical modeling. Earthq Eng Struct Dyn 41:1845–1860

Yeoh OH (1993) Some forms of the strain energy function for rubber. Rubber Chem Technol 66(5):754–771

Chapter 8
Experimental Methods and Activities in Support of Earthquake Engineering

Stathis N. Bousias

Abstract The role of experimental methods as an indispensable element supporting earthquake engineering research activities, is presented. Static, dynamic and pseudodynamic (or hybrid simulation) methods have accompanied and, in some cases, triggered the advancements seen during the last decades in earthquake engineering. Structural response, be it for the design of new structures or for the assessment and retrofitting of existing ones, has been thoroughly studied with the aid of experimentation that produced the necessary volume of data to calibrate numerical models and support code provisions. In the following, through examples of experimental studies employing a variety of testing methods it is shown that experimentation is the sine qua non ingredient for the advancement of earthquake engineering research and practice.

Keywords Testing methods · Structural testing · Pseudodynamic testing

8.1 Introduction

Engineering has always been based on observation and experimentation. For structural engineering, and particularly earthquake engineering, this approach is still one of the cornerstones of the advancements realized, as evidenced by the effort paid in constructing new large scale testing installations. The empiricism that characterized it initially gave way to rigorous methods, which became more and more powerful owing to technological advancements. In a broad characterization (due to the nature of the seismic action the rate of loading is used as a criterion) three are the main thrust areas of experimentation in earthquake engineering research: static, pseudodynamic and fully dynamic testing. With each method providing data for certain aspects of structural response and with certain pros and

S.N. Bousias (✉)
Structures Laboratory, Department of Civil Engineering,
University of Patras, 265 00 Patras, Greece
e-mail: sbousias@upatras.gr

© Springer International Publishing AG 2017   139
A.G. Sextos and G.D. Manolis (eds.), *Dynamic Response of Infrastructure
to Environmentally Induced Loads*, Lecture Notes in Civil Engineering 2,
DOI 10.1007/978-3-319-56136-3_8

cons with respect to the others, it is natural that all three are still been used. Although technological progress favours pseudodynamic or dynamic testing methods much more that it does for static testing, it works against the diffusion the former methods enjoy and the cost associated with their implementation.

In the following, only the basic characteristics of each testing method are presented with emphasis being placed more on their contribution to the advancement of earthquake engineering—appropriate examples are employed through the multiannual experimental activity at the Structures Laboratory of the University of Patras.

8.2   Static Testing

Testing of structures under a slow deformation/force rate has been extensively used in the past: low-flow capacity static actuators are employed to apply a pre-determined pattern of forces or displacements on full- or nearly full-scale specimens. Although the requirements for the loading equipment can be easily met, the test set-up may be complicated, especially when structural components (e.g. columns) or subassemblies (e.g. joints) are being tested and the actual boundary conditions and loading have to be met. Static testing allows for stop/restart of the process, permits easy observation of crack-pattern development and posses no special requirements for sensor and data acquisition selection. Static testing has been the main horsepower behind the development and calibration of the majority of detailed or of member-type models for the nonlinear response analysis of structures subjected to seismic excitation. Also, in both the early and later stages of the research towards developing retrofitting techniques for damaged or sub-standard structures, it constituted the main tool for studying the effectiveness of a wide variety of repair/retrofitting measures—a examples of such application is cited next.

8.2.1   Structural Repair/Retrofitting

Old, substandard concrete structures often suffer both from deficiencies in member strength and deformation capacity, as a result of one or more characteristics of reinforcement and reinforcing detailing:

- smooth reinforcing bars in the longitudinal direction,
- hook-spliced smooth bars,
- inadequate length of lap splicing of deformed reinforcing bars,
- sparsely placed stirrups not providing adequate confinement and/or shear strength, and
- reinforcement corrosion

In addition to the above, reinforcement corrosion not only reduces member strength due to steel area loss, but it also makes bars more susceptible to buckling, reduces steel ductility while affecting negatively bond and anchorage. As a result of all/some of the above parameters existing structures are underperforming with respect to current code requirements. Because columns are the elements encompassing most of the above deficiencies and with the most decisive contribution to structural response, retrofitting measures aim at re-instating column flexural capacity mainly through jacketing with reinforced concrete or via fiber reinforced polymers. A series of issues are encountered when designing for the retrofitting:

- Thickness of the RC jacketing
- Methods to connect old concrete to the RC jacket

  - Roughening the surface
  - Provision of dowels and inserts
  - Combination of surface roughening with dowels
  - No special measures

- Type of fiber material (carbon or glass) of the polymer jacket
- Influence of the aspect ratio of the member cross-section
- Number of layers per type of fiber polymer material
- Height of application of the FRP jacket

It was only after extensive experimental research, plus the data produced, that the basis for the design of the interventions shifted from purely empirical to more rigorous, now incorporated in the present codes. The way to study experimentally the influence of a large number of parameters is to test a member rather that a structure (Fig. 8.1), as the manageable size of the specimen along with the static application of loading simplifies the testing setup and reduces the cost of the campaign. In the testing campaigns by Bousias et al. (2004, 2006, 2007a, b) more than 50 columns were tested (smooth or deformed bars, lap-spliced or continuous longitudinal reinforcement, with and without damage before retrofitting and with/without bar corrosion) with many important findings and with the results been used for the calibration of analytical models (e.g. Thermou et al. 2014) and the development of relevant code provisions (e.g. EN1998-3). In summary, it was found that:

- Lap-splicing of deformed bars for at least 45-bar diameters does not significantly reduce cyclic deformation capacity (compared to a column with continuous bars)—the opposite is true for lapping by as little as 15-bar diameters (appreciable reduction of flexural resistance, rapid post-peak strength and stiffness degradation and low energy dissipation capacity). Concrete jackets seem quite effective in remedying the above adverse effects, even for very short lap lengths (the reduced hysteretic energy dissipation observed in the original column is this case retained in the jacketed one). FRP-wrapping leads to significant improvements in flexural resistance, ultimate deformation and energy dissipation—the improvement is reduced with the reduction of lap length.

Fig. 8.1 Static testing of RC columns: original configuration (*top left*); jacketed with RC (*top right*); measures to improve interface conditions (*middle row*); jacketed with FRPs/FRP rupture (*bottom*)

Tests revealed that there is a limit to the improvement brought by the FRP wrapping, with the lower end being the 15 bar-diameters lap length for which FRP jacketing cannot sufficiently remove the deficiency caused on force capacity and energy dissipation.

- Non-seismically designed columns with smooth (plain) vertical bars have low deformation and energy dissipation capacity under cyclic loading, which is however not impaired much further by lap splicing—at least for lapping as short as 15 bar-diameters. A lap length of at least 15-bar diameters seems to sufficiently supplement the 180° hooks for the transfer of forces. Retrofitting such members with RC jacketing increases their deformation capacity to levels sufficient for earthquake resistance, irrespectively of the presence and length of lap splicing. Lack of positive measures to connect the old concrete to the jacket (surface roughening, dowels, inserts) does not adversely affect the lateral load resistance, deformation capacity or energy dissipation of the jacketed column. On the other hand, two layers of FRP-wrapping of columns with smooth bars and hooked ends suffice, irrespectively of the lap length. However, a decrease in lap length seems to reduce energy dissipation in the FRP-retrofitted column. Overall, FRP wrapping of just the plastic hinge and any splice region was shown to be more effective than concrete jacketing in enhancing the deformation and energy dissipation capacity of old-type columns having smooth bars with or without lap-splicing. One notable outcome of the research was that FRP-wrapping of an end region of the member length equal to 1.2 times the section depth cannot preclude plastic hinging and early member failure outside the FRP-wrapped length of the column.
- Reinforcement corrosion causes a gradual loss of lateral and axial load resistance during cycling that, ultimately, leads to failure. Deformation capacity and hysteretic response improve considerably when non-ductile regions are encased in continuous FRP jackets with either carbon or glass fibers. FRP-retrofitted columns maintain practically constant lateral force capacity up to ultimate deformation, but lose it abruptly when they fail explosively by fracture of the FRP wrap. Parametric tests showed that confinement is controlled by the FRP extensional stiffness: employing glass or carbon fibers while maintaining the same extensional stiffness of the FRP jacket in the circumferential direction, leads to about the same performance. A point of practical usefulness raised from the results was that although FRP wrapping significantly improves seismic performance of columns which suffer from both lack of seismic detailing and of reinforcement corrosion, such corrosion materially reduces the effectiveness of FRP wraps as a strengthening measure.

Static testing has, however, not only been used for testing members—important information can be derived also regarding the global structural response, e.g. in case of a particular type of intervention aiming at modifying the global structural response, as, for example, the addition of stiffening elements (e.g. diagonal braces or shear walls). As most building structures are masonry infilled, the possibility of combining the high initial stiffness and strength of masonry with the higher

Fig. 8.2 Testing of 3-storey specimen retrofitted with TRM-jacketing

deformability and sustainability of cyclic lateral load capacity offered by innovative materials so as to yield a more reliable structure response, is worth exploring.

In the experimental study by Koutas et al. (2014) seismic retrofitting of nearly full-scale 3-storey masonry infilled frames employing non-conventional materials and techniques, was investigated. The application of textile-reinforced mortar (TRM) as externally bonded reinforcement in combination with special anchorage details was examined on an as-built and a retrofitted, 2:3 scale RC frame, subjected to in-plane loading (Fig. 8.2). In this case static application of a triangular force pattern was selected, so as a 'clear' response can be obtained (useful in developing relevant numerical models) as compared to that of shake table or hybrid testing (sometimes obscured by the particularities of the seismic motion and other secondary effects). Static testing allows for a better description of the sequence of development of the cracking pattern on masonry a particularly useful element for characterizing the contribution of the textile-reinforced mortar jacketing selected for retrofitting. The application of TRM over the entire surface of infills was

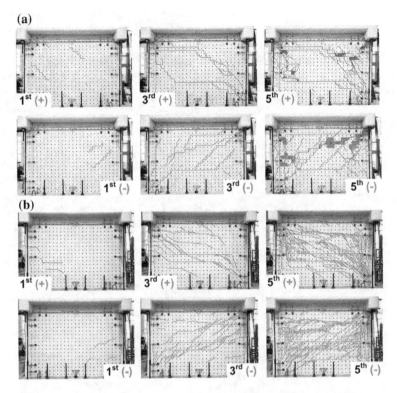

Fig. 8.3 Bottom storey crack pattern: **a** control specimen; **b** retrofitted specimen

supplemented with an adequate infill-frame connection materialized via custom-fabricated textile-based anchors. The test results verified the enhancement of lateral strength and deformation capacity offered by retrofitting; an approximate 56% increase in lateral strength, 52% increase in deformation capacity and 22.5% increase in energy consumption. In addition, the scheme succeeded in suppressing pre-emptive column shear failure observed in the control specimen (Fig. 8.3).

Between the two extremes of member or full structure testing under static conditions, structural subassemblies may also be tested to provide experimental evidence on the response of structural systems, with one such example being the study of structurally resilient systems which strive to fulfill (the grossly prescriptive) code provisions. In a series of ongoing tests by Stathas et al. (2015a), the column-collapse scenario and how alternative structural systems may be of assistance in coping with the resulting disproportionate consequences, was examined by testing subassemblies comprising of the elements neighboring in the vicinity of the removed vertical element. The specimens tested represented (Fig. 8.4):

- A two-span portion of slab with appropriate boundary conditions, tested under vertical loading: two specimens, one reinforced according to currently accepted practices and one with a different detailing, were tested.

146   S.N. Bousias

Fig. 8.4 Static testing of subassemblies for resilient systems under column-removal scenario: subassembly comprising slab portion (*top*); beam subassembly (*middle*); full scale frame tested under cyclic lateral loading (*bottom*)

- A two-span beam framing on a collapsed column tested under vertical loading: specimens included a beam with conventional reinforcing pattern and a second with an alternative system incorporating straight, unbonded axial prestressing along with dry-joints at beam-column interface and unbonded beam longitudinal reinforcement along the whole column width.
- A two-story frame tested under lateral cyclic loading; the beams of the frame were constructed according to the alternative system of dry joints and axial

prestressing tested earlier for the column removal scenario. The case of rocking of the foundation was also examined through a special mechanism at the base of all column foundations.

8.3 Shake Table Testing

The closest-to-reality reproduction of the seismic response of structures can be obtained via shake table testing: a model of the structure placed on the stiff platen is moved in real time via an ensemble of dynamic actuators, following a predefined natural or artificial accelerogram. In shake table testing the advantage of full replication of the actual effects of the earthquake excitation is, in general, not commensurate to the cost, complexity of the infrastructure and the need for dimensionally scaling the specimen under test. Testing with this type of experimental device received wider attention only after the mid-60s, following the development of appropriate power transmission devices—for a historical note on the development of shaking tables the interested reader is referred to Severn (2011).

The complex interaction phenomena developing during real time testing and the variability of the response both in amplitude and frequency (partially owing to the particularities of the seismic motion), do not lend the results of shake table testing amenable to direct use for building numerical models for the nonlinear response of structures—instead, the results of static testing have been mostly employed for this purpose. Nevertheless, it is in many instances that it is realized that shake table testing is simply irreplaceable: rate-sensitive devices (e.g. dampers) incorporated in buildings for controlling the level of displacements may, of course, be tested individually, but for obtaining the total response they have to be tested on a shake table together with the rest of the structure. Another problem that can only be studied via shake tables is the response of structures allowed to rock by-design. Makris and Vassiliou (2013) have elaborated theoretically the issue of rocking to prove that, if bridge piers that support the deck were allowed to rock atop their pile caps, two major advantages would result: (a) rocking systems exhibit negative stiffnesses; thus, their response neither amplifies nor resonates from any frequency content of the input ground motion; and (b) re-centering is achieved through gravity, an asset that is available for free. In an ongoing testing campaign undertaken by Makris et al. (2015) involves testing a 1:3 free-standing replica of a circular bridge pier (Fig. 8.5, left) under an ensemble of carefully selected long-period pulse like ground motions. The contribution of gravity in re-centering is examined via a pair of free-standing columns capped with a freely supported rigid beam (Fig. 8.5, right). Preliminary experimental results show appreciable agreement with the theoretical predictions.

Fig. 8.5 Shake table testing of rocking structure: free-standing single pier (*left*) and freely supported rigid beam capping a pair of columns

8.4 Pseudodynamic Testing (Hybrid Simulation)

The roots of the pseudodynamic (PsD) testing method date back in 1969, when it was first proposed by Hakuno et al. (1969), and it was first applied by Takanashi et al. (1975). By combining the simplicity, reasonable cost and regular testing equipment with the realistic excitation employed in shake tables, the method has seen appreciable diffusion, as well as many promising extensions. By leveraging equipment regularly found in structural laboratories, the method has proven very competitive against costly shaking table testing, especially when testing in full scale is concerned. With the exception, maybe, of its application to testing rate-dependent materials/devices and to distributed-mass systems, intensive research during the last 15 years has successfully treated many of its initial weaknesses. Furthermore, the extension of the method on the basis of the concept of substructuring (i.e. the discretization of selected parts of the structure under test into actively interacting physical and numerical substructures forming a, so-called, "hybrid model"), has resulted in a valuable tool for testing structures of such large size that could not otherwise be accommodated in present-day laboratories (e.g. bridges). It was exactly the concept of substructuring that opened up new possibilities and allowed for innovative applications of the method, such as the geographically distributed sub-structured testing and the real-time sub-structured testing. New, open-source software tools have also been developed to facilitate the co-ordination and execution of distributed hybrid simulations. To encompass these developments the term "hybrid simulation" was coined in recent years, principally by the Network for Earthquake Engineering Simulation (NEES, http://www.nees.org) the members of which provided considerable impetus to the method. A brief description of the application of the classic PsD method as well as of some of its extensions, is provided next.

8.4.1 Classic PsD Testing

The classic PsD testing (Molina et al. 1999) involves the determination of the structural response through the numerical integration of the equation of motion, with the difference that the information regarding the state of structural stiffness is being experimentally obtained. The whole process evolves in an expanded time scale such that each step of the discretized input motion lasts at least one-hundred times longer. As long as there is no conflict between this time expansion and the material behaviour (e.g. rate-sensitivity), the method yields accurate results and allows for the execution of complicated experiments with an appreciable number of actuators being involved. Testing of a two-story, torsionally-unbalanced reinforced concrete structure in nearly full-scale, is an example of the potential of the method. Along the lines of the tests by Molina et al. (1999), Bousias et al. (2007a, b) tested a 4-Dof structure with floor deformations (displacement and rotation) applied via a pair of actuators per slab. The equations of motion are solved in a separate machine and communicated to the controllers via a local network. Several structures of the same geometric configuration but with different characteristics were tested: as-built structure (including detailing typical of substandard structures), structure retrofitted with fiber reinforced polymer plies (FRP), structure with symmetry reinstated via reinforced concrete jackets and structure with masonry infilling at first storey to yield a soft-story configuration (Fig. 8.6). The experimental campaign allowed to investigate the nonlinear response of the as built structure, as well as to verify the effectiveness of different approaches (intervention to ductility and/or stiffness) regarding the retrofitting measures that can be considered for such structures. More details can be found in Bousias et al. (2007a, b).

Retrofitting at the global level has also been examined though PsD testing: instead of refurbishing individual members with increased ductility, strength and/or stiffness, structure-level retrofitting techniques may be sought through the addition of steel bracing or of new reaction walls. The possibility of replacing the more demanding option of addition of new shear walls with RC infilling of consecutive

Fig. 8.6 Pseudodynamic testing of substandard structures: structure as-built (*left*), RC-jacketed structure (*middle*) and pilotis-type structure retrofitted with FRP (*right*)

Fig. 8.7 Pseudodynamic testing of a 4-story RC infilled frame

bays along-height, was investigated by Strepelias et al. (2013). A four-story frame
with RC infilling (Fig. 8.7) was tested pseudodynamically, to investigate two—
remarkably different in cost—reinforcing schemes for connecting the new concrete
to the surrounding concrete members.

8.4.2 Substructured PsD Testing

The introduction of substructuring in the pseudodynamic method and the development of software tools for performing hybrid simulation have paved the way for a
new era in structural testing. The capability of judiciously discrediting the structure
into modules and solving for the whole structure employing information for the
state of each module being provided either from appropriate numerical models
(numerical substructures) or from physically subjecting the module (physical substructure) to the deformations commanded by its response as part of the whole
structure, offers flexibility, economy and the means to perform more focused testing
at large scale. Consider, for example, the case of a bridge structure (Fig. 8.8): its
response is the combined contribution of the pier, the elastomeric bearings at end

supports and the interaction of the bridge with the underlying soil. For some of the structural parts (e.g. bearings) reliable numerical models suffice for obtaining their response—the same does not hold for the response of the pier. Thus, a physical model of the pier respecting the actual boundary conditions in the real structure can be constructed (Fig. 8.8), with the rest structural parts being treated analytically (Stathas et al. 2015b). The information produced by each module is communicated to the coordinating software to produce command displacements to be communicated at and imposed to each module during the following time step. Dedicated

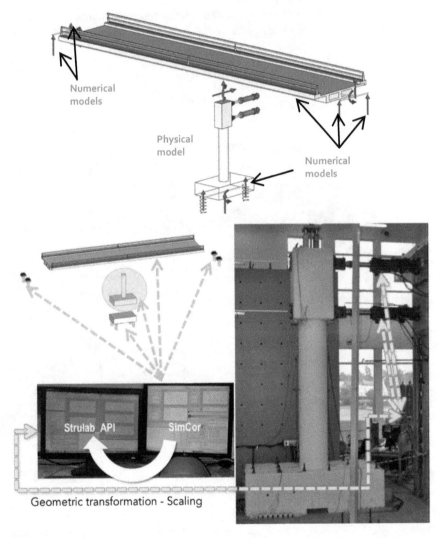

Fig. 8.8 Sub-structured pseudodynamic testing of a bridge: structural discritization (*top*); module communication architecture (*bottom*)

software needs to be developed to handle issues regarding scaling, geometric transformations and limit checks.

8.4.3 Geographically Distributed Hybrid Simulation

Hybrid simulation is rapidly expanding testing method: it is based on sub-structuring the structure in question and treating each sub-structure either analytically or experimentally. Owing to its versatility, capacity to cope with member, subassembly or full-scale structural testing (pseudodynamic or real-time) and its flexibility in combining the experimental/computational potential of

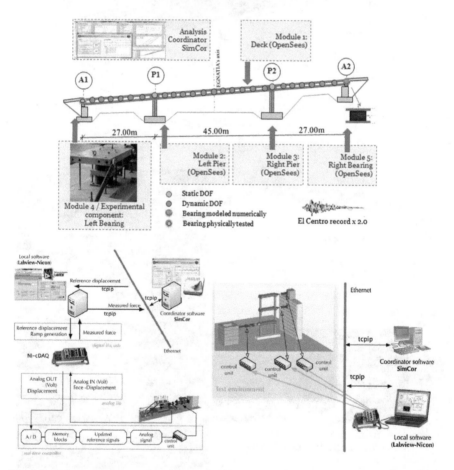

Fig. 8.9 Hybrid simulation of multi-span bridge: system architecture (*top*); communication scheme via analog input (*bottom left*), and com communication scheme via digital input (*bottom right*)

individual laboratories, the method has been employed in many testing campaigns, especially within the NEES project.

In all cases, a simulation-controlling component ("coordinator") is entrusted with the time-stepping integration algorithm and the coordination of the communication between substructures. Complying to the established terminology (Nakata et al. 2014), two configurations may be employed: the coordinating component is a designated software (e.g. Kwon et al. 2005, 2008) communicating with external finite element codes and physical testing for the numerical and experimental substructures, respectively. In a second approach (OpenFresco/Opensees platform, Schellenberg et al. 2009) the analysis of the numerical substructures is performed within the finite element software and the only network communication required is that with the laboratory-tested component(s). An important issue that needs to be resolved regardless of the approach opted for concerns the communication of computed deformations to the laboratory controller. The options available for realizing a two-way communication path between the simulation coordination software and the laboratory control system (Fig. 8.9), was investigated and implemented by Bousias et al. (2014) for the example of a multispan bridge that was tested in a geographically distributed substructured configuration employing international partners.

8.5   Outreach of Experimental Results

Given the investment in cost, time, technology and know-how involved in large scale testing—be it static, pseudodynamic or fully dynamic—maximization of the usefulness and diffusion of test results should be sought. Nevertheless, it is common practice that the data remains with the laboratory that produced it, thus limiting its exploitation and widespread usage. Diffusion of the test results is one of the facets of data exploitation, with a second one being linked to the maximization of the information that can be extracted from the experimental results. Last, but not least, should physical access, or even virtual presence, of interested researchers be provided to the experimental process, the experience of testing at large scale could be spread and synergy between geographically distant teams could be facilitated.

With the NEES consortium been the first to depart towards this direction and developing the necessary tools, the increase in the value of the experimental results has been based on a two-fold approach: integration of data acquisition systems with video systems and development of databases (centralized or de-centralized) for collecting, storing and diffusing structured experimental data. The integration of data acquisition systems with video systems is based on the time-stamping of data: data acquired from the experimental equipment are associated time-wise to photos and video taken during testing. This allows, for example, to correlate a specific event, a sequence of events or an individual local phenomena during the response (e.g.cracking, bar fracture, bar buckling, etc.) to the exact value of all measured data at the instant of the event. If, in addition, the time-stamped data was available to

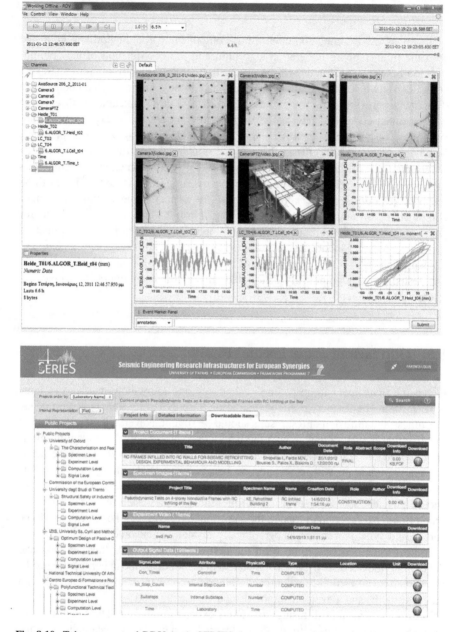

Fig. 8.10 Telepresence tool RDV (*top*); SERIES data access portal (*bottom*)

distant researchers in real-time, or via an archive file, the scope of data diffusion and 'telepresence' is fulfilled. All this has been possible with the aid of dedicated software (RDV, Real-time data viewer) whose use allows live or archived time-synchronized data to be viewed, analyzed or even played back (Fig. 8.10).

The issue of data curation has been treated in two different ways: the one followed by NEES consortium comprised of a database seen as a centrally-managed depository which collects and archives data from a group of laboratories. Given certain disadvantages of this scheme, the SERIES virtual database (Bosi et al. 2015; Lamata et al. 2015) is composed of an ensemble of local databases residing at the laboratories producing the data. A central portal (http://www.dap.series.upatras.gr) has been created (Fig. 8.10) providing users centralised access to the individual database nodes distributed over the network (i.e. to each laboratory database). Local laboratory databases are fed with the data the laboratory volunteers to share and the central portal queries each local database to identify existing data which fulfil the request of external users. This architecture preserves the ownership of the data and avoids the need to store and manage centrally voluminous data. A purpose--developed data exchange format has been developed so that the database archives not only numeric data but also metadata and reports in addition time-stamped video and photos.

8.6 Conclusions

Experimental testing has offered a long-standing contribution to the achievements of earthquake engineering research, furnishing the data necessary to cover gaps in a very demanding discipline. Laboratory infrastructures, employing one or more of the available methods suitable for the problem at hand, have taken full advantage of developments in technology to offer an as much as possible accurate representation of the highly nonlinear response of structural systems subjected to seismic action. Testing has been of assistance in the whole spectrum of earthquake engineering research activities: the design of seismic resistant structures, the development of experimentally verified code provisions, the assessment and retrofitting of structures without seismic detailing, the problem of soil-structure interaction and many others so on. For the foreseeable future, experimental testing is expected to continue providing valuable data, not only through the construction of more powerful and flexible infrastructures with unique characteristics, but also through the contribution of existing laboratories of any size, operating individually or through technology-driven synergistic ways.

Acknowledgements The contribution of all students and research associates who participated in the experimental campaigns in the last 20 years at Structures Laboratory of the University of Patras, is greatly acknowledged. The studies were funded by several agencies through research projects, the majority of which were coordinated by Prof. M. Fardis.

References

Bosi A, Kotinas I, Lamata Martínez, Bousias S, Chazelas JL, Dietz M, Hasan MR, Madabhusi SPG, Prota A, Blakeborough T, Pegon P (2015) Ch. 4: The SERIES virtual database: exchange data format and local/central databases. In: Taucer F, Fardis M (eds) Earthquake engineering research infrastructures, vol 35. Springer International Publishing, pp 31–48. ISBN 978-3-319-10136-1

Bousias SN, Triantafillou TC, Fardis MN, Spathis L-A, O'Regan B (2004) Fiber-reinforced polymer retrofitting of rectangular RC columns with or without corrosion. ACI Struct J 101(4): 512–520

Bousias SN, Spathis L-A, Fardis MN (2006) Concrete or FRP jacketing of columns with lap splices for seismic rehabilitation. J Adv Concr Technol 4(3):1–14

Bousias SN, Spathis L-A, Fardis MN (2007a) Seismic retrofitting of columns with lap-spliced smooth bars through frp or concrete jackets. J Earthq Eng 11:653–674

Bousias SN, Fardis MN, Spathis L-A, Kosmopoulos A (2007b) Pseudodynamic response of torsionally unbalanced 2-story test structure. J Earthq Eng Struct Dyn 36:1065–1087

Bousias S, Kwon O-S, Evangeliou N, Sextos A (2014) Implementation issues in distributed hybrid simulation. In: Proceedings of the 6th world conference of structural control and monitoring, Barcelona

Hakuno M, Shidawara M, Hara T (1969) Dynamic destructive test of a cantilever beam controlled by an analog-computer. Trans Jpn Soc Civil Eng 1–9

Koutas L, Bousias S, Triantafillou T (2014) Seismic strengthening of masonry-infilled RC frames with TRM: experimental study. ASCE J Compos Constr. doi:10.1061/(ASCE)CC.1943-5614. 0000507

Kwon O-S, Nakata N, Elnashai A, Spencer B (2005) A framework for multi-site distributed simulation. J Earthq Eng 9(5):741–753

Kwon O, Elnashai AS, Spencer BF (2008) A framework for distributed analytical and hybrid simulations. Struct Eng Mech 30(3):331–350

Lamata IM, Ioannidis I, Fidas C, Williams M, Pierre Pegon (2015) Ch. 3: The SERIES virtual database: architecture and implementation. In: Taucer F, Fardis M (eds) Earthquake engineering research infrastructures, vol 35. Experimental research in earthquake engineering. Springer International Publishing, pp 31–48. ISBN 978-3-319-10136-1

Makris N, Vassiliou MF (2013) Planar rocking response and stability analysis of an array of free-standing columns capped with a freely supported rigid beam. Earthq Eng Struct Dyn 42 (3):431–449

Makris N, Alexakis C, Kampas G, Strepelias E, Bousias S (2015) SeismoRockBridge project: seismic protection of bridges via rocking of their piers which re-center with gravity—learning from ancient free-standing temples: experimental and theoretical studies. Report to the General Secretariat for Research (in Greek)

Molina FJ, Verzeletti G, Magonette G, Buchet P, Géradin M (1999) Bi-directional pseudodynamic test of a full-size three-storey building. Earthq Eng Struct Dyn 28:1541–1565

Nakata N, Dyke S, Zhang J, Mosqueda G, Shao X, Mahmoud H, Head M, Erwin M, Bletzinger M, Marshall GA, Ou G, Song C (2014) Hybrid simulation primer and dictionary. Network for earthquake engineering simulation, NEES

Schellenberg AH, Mahin SA, Fenves GL (2009) Advanced implementation of hybrid simulation. PEER Report 2009/104

Severn RT (2011) The development of shaking tables—A historical note. Earthq Eng Struct Dyn 40:195–213. doi: 10.1002/eqe.1015

Stathas N, Palios X, Fardis M, Bousias S, Skafida S, Digenis S (2015a) Paradigm for resilient concrete infrastructures to extreme natural and man-made threats. Report to the General Secretariat for Research (GSRT)

Stathas N, Skafida S, Bousias S, Fardis M, Digenis S, Palios X (2015b) Hybrid simulation of bridge pier uplifting. Bull Earthq Eng (Special Issue: Large scale and on-site structural testing for seismic performance assessment). doi:10.1007/s10518-015-9822-2

Strepelias E, Palios X, Bousias SN, Fardis MN (2013) Experimental investigation of concrete frames infilled with RC for seismic rehabilitation. ASCE J Struct Eng. doi:10.1061/(ASCE)ST. 1943-541X.0000817

Takanashi K, Udagawa K, Seki M, Okada T, Tanaka H (1975) Non-linear earthquake response analysis of structures by a computer-actuator on-line system (Part 1 detail of the system). Transcript of the Architectural Institute of Japan

Thermou GE, Papanikolaou VK, Kappos AJ (2014) Flexural behaviour of reinforced concrete jacketed columns under reversed cyclic loading. Eng Struct 76:270–282

Chapter 9
Time Reversal and Imaging for Structures

C.G. Panagiotopoulos, Y. Petromichelakis and C. Tsogka

Abstract We present a numerical implementation of the time-reversal (TR) process in the framework of structural health monitoring. In this setting, TR can be used for localizing shocks on structures, as well as, for detecting and localizing defects and areas which have suffered damage. In particular, the present study is focused on beam assemblies, typically utilized for simulating structures of civil engineering interest. For that purpose, Timoshenko's beam theory is employed since it is more adequate for describing higher-frequency phenomena. The numerical procedure is explained in detail and the capabilities of the proposed methodology are illustrated with several numerical results.

Keywords Time-reversal · Structural health monitoring · Finite element method · Imaging

9.1 Introduction

Wave propagation in structural components is a topic which appears to be important in numerous cases of engineering interest (Graff 1975). There are many structures that actually act as, or can be approximated by, assemblings of one-dimensional beam components. In many cases bridges fall in this category, with the typical example

C.G. Panagiotopoulos · Y. Petromichelakis · C. Tsogka (✉)
Institute of Applied & Computational Mathematics, Foundation for Research and Technology Hellas, Heraklion, Greece
e-mail: tsogka@uoc.gr

C.G. Panagiotopoulos
e-mail: pchr@iacm.forth.gr

Y. Petromichelakis
e-mail: yiannispetro@gmail.com

C. Tsogka
Department of Mathematics & Applied Mathematics, University of Crete, Heraklion, Greece

© Springer International Publishing AG 2017
A.G. Sextos and G.D. Manolis (eds.), *Dynamic Response of Infrastructure to Environmentally Induced Loads*, Lecture Notes in Civil Engineering 2, DOI 10.1007/978-3-319-56136-3_9

being that of truss bridges. We consider here the problem of source shock or defect localization in such structures.

The problem of detection and localization, of defects and areas suffered damage, based on recordings at limited number of spatial points, falls into the category of inverse problems which are usually ill-posed and hard to solve. A computational tool for solving a class of inverse wave (and/or vibration) problems is the time reversal (TR) technique which was originally introduced in Prada et al. (1995) as a physical process. The principal idea behind TR is to back-propagate the recorded signals but reversed in time. TR consists of two steps: a forward and a backward propagation one. In the forward propagation step, waves are emitted from some source and travel through the medium. During this step the wave-field is being recorded by one or more receivers. In the backward step, the previously recorded signals are reversed in time and they are rebroadcasted from their respective receiver positions. Wave paths that were traversed in the forward propagation step are now reproduced in the backward one Anderson et al. (2008). Ideal conditions for the time reversal process would be those corresponding to the case where receivers fill the whole medium (or its entire boundary), recording the field and its derivatives Fink and Prada (2001), without any noise Givoli (2014). The time *reversibility* is based on the spatial *reciprocity* (symmetry in engineering) and the time reversal *invariance* (under the transformation $t \rightarrow -t$) of linear wave equations. Because of the time-reversibility of the wave equation this procedure leads to a wave that refocuses at the region of the source. A defect or damaged area, can be understood to act as a secondary source and therefore TR can be used for its localization as well. The time reversal technique has been recognized in recent years, because of its robustness and simplicity, as a quite appealing approach with application to numerous disciplines. Interested reader is referred to review articles that exist in the literature with extended presentation of the TR technique and its applications (Yavuz and Teixeira 2009; Givoli 2014).

We consider here structures that can be modelled by one-dimensional computational domains such as truss, beam elements, frame structures, etc. For this set-up we develop and test a time reversal procedure that allows for the location of sources, as well as, damaged areas in the structure. Note that in the case of damage our approach relies on the scattered field which is obtained by subtracting the wave-field in the healthy structure from the field in the damaged structure. A migration or back-propagation imaging technique that allows for damage localization is also described. It is worth mentioning that our approach, although presented in the one-dimensional setting, it can be easily generalized to two- and three-dimensional domains.

9.1.1   Longitudinal Waves in Thin Rods

We first consider a simple one-dimensional model that describes propagation of axial waves in a thin homogeneous rod (Graff 1975). We denote by l, the length of the rod and by A, the cross-section area, which is assumed constant over length. The material properties are described by the elasticity modulus, E, and the mass density, ρ. Let us also denote the longitudinal displacement by $u(x, t)$, the stress field as $\sigma(x, t)$ and the body force as $q(x, t)$.

Assuming linear elastic behaviour and following Hooke's law $\sigma = E\varepsilon$, we obtain the equation of motion,

$$A E \frac{\partial^2 u}{\partial x^2}(x, t) - A \rho \frac{\partial^2 u}{\partial t^2}(x, t) = q(x, t). \tag{9.1}$$

Here ε is the axial strain defined by $\varepsilon = \frac{\partial u}{\partial x}$. We add to (9.1) initial,

$$u(x, 0) = u_0(x), \qquad \frac{\partial u}{\partial t}(x, 0) = \dot{u}_0(x), \tag{9.2}$$

and boundary conditions

$$u(x, t) = g(t), \quad \forall x \in \Gamma_D, \qquad \frac{\partial u}{\partial x}(x, t) = h(t), \qquad \forall x \in \Gamma_N. \tag{9.3}$$

Note here that we use the notation Γ_D for the boundary part where displacements, or the generalized degrees of freedom are known. The boundary part Γ_N where spatial derivatives or forces, compatible to the degrees of freedom, are imposed. Finally, we may use Ω for describing the medium itself. In the absence of body forces, that is $q(x, t) = 0$, Eq. (9.1) reduces to the well known wave equation

$$\frac{\partial^2 u}{\partial x^2} = \frac{1}{c_0^2} \frac{\partial^2 u}{\partial t^2}, \tag{9.4}$$

with $c_0 = \sqrt{\frac{E}{\rho}}$, the wave propagation velocity.

The general solution for Eq. (9.1), under various boundary and initial conditions, as well as, external loading, can be obtained considering an eigenfunction expansion. Alternatively, an integral transformation approach (Laplace, Fourier) or the Green's function method can be followed (Graff 1975).

Waves incident on the boundary of the domain are reflected in a manner that depends on the nature of the boundary condition as given from Eq. (9.3), or an even more complex (e.g. mixed type) expression. Fixed displacement ($u(x, t) = 0$, $x \in \Gamma_D$) and free end ($\frac{\partial u}{\partial x}(x, t) = 0$, $x \in \Gamma_N$) boundary conditions are the two most common cases encountered in applications.

As can be seen in Fig. 9.1, for the case of fixed end boundary the original pulse changes sign after reflection, while for the case of free end, shown in Fig. 9.2 the spatial derivative of the original pulse and thus the stress changes sign.

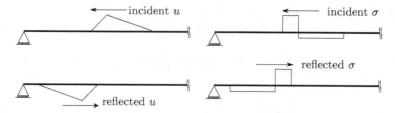

Fig. 9.1 Reflected displacement and stress of incident wave on a fixed end

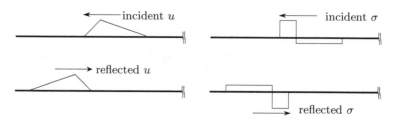

Fig. 9.2 Reflected displacement and stress of incident wave on a free end

9.1.2 Flexural Waves in Thin Rods

A more general model that comprises flexural wave propagation can be obtained using the Bernoulli-Euler beam theory. The resulting system of equations is dispersive and allows for infinite wave propagation speed (Graff 1975) which is certainly unrealistic and contradicts the theory of elastodynamics. Later, Rayleigh solved this problem partially by introducing rotary inertia. We employ here instead Timoshenko's beam theory (Timoshenko 1921, 1922) which includes effects of shear and rotary inertia, that are neglected in the previously mentioned Bernoulli-Euler theory. To be more specific, Timoshenko beam incorporates the following two refinements over the Bernoulli-Euler model,

- For both statics and dynamics, plane sections remain plane but not necessarily normal to the deflected midsurface. This assumption allows the averaged shear distortion to be included in both strain and kinetic energy.
- In dynamics, the rotary inertia is included in the kinetic energy,

$$\kappa GA \left(\frac{\partial \psi}{\partial x} - \frac{\partial^2 v}{\partial x^2} \right) + \rho A \frac{\partial^2 v}{\partial t^2} = w(x, t), \tag{9.5a}$$

$$\kappa GA \left(\frac{\partial v}{\partial x} - \psi \right) + EI \frac{\partial^2 \psi}{\partial x^2} - \rho I \frac{\partial^2 \psi}{\partial t^2} = \mu(x, t). \tag{9.5b}$$

Here v the transverse displacement and ψ measures the slope of the cross-section due to bending. An additional contribution γ_0 due to shearing effect is included since,

$$\frac{\partial v}{\partial x} = \psi + \gamma_0. \qquad (9.6)$$

External excitation is considered as the prescribed distributed lateral load $w(x,t)$ on the beam and an action of prescribed distributed moments $\mu(x,t)$. Here ρ is the mass density, A is the cross-section area, E is the modulus of elasticity, G is the modulus of rigidity, I is the moment of inertia of the cross-section, and finally κ is a coefficient that depends on the shape of the cross-section (Cowper 1966). Timoshenko's beam theory assumes two modes of deformation and Eq. (9.5) represents the physical coupling that occurs between them. Wave velocities due to Timoshenko's beam theory are bounded at large wavenumbers having two asymptotic limits, because of the two basic modes of motion (shear and bending) (Fung 1965), that actually are $c_1^2 = \dfrac{E}{\rho}$ and $c_2^2 = \dfrac{\kappa G}{\rho}$.

9.1.3 Assembly of Structural Members

The most convenient way to introduce more complicated structures, consisting of a collection of arbitrarily oriented one-dimensional components, is by matrix representation. A suitable approach is the finite element method (FEM), for which stiffness and mass matrix for axial deformation, as well as, for transverse displacement and slope according to Timoshenko beam, are available (Przemieniecki 1968). This is illustrated in Fig. 9.3, where a junction of three distinct one-dimensional domains that meet each other, is shown. Each of the structural components, is assumed to be capable of bearing axial as well as bending loading. Axial and transverse-rotational modes are uncoupled, as has been explained earlier (9.5), on the local reference coordinate system. However, on the global reference coordinate system, and because of

Fig. 9.3 A junction of three coupled beams (1D elastic domains), of different characteristics (cross-section, material), with corresponding propagating waves

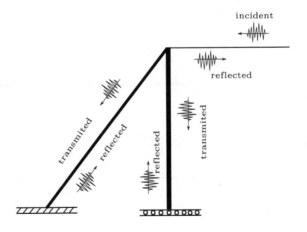

the interaction at the junctions of the assembly, coupling takes place. At the junction, equilibrium of forces and continuity of displacements is forced, while additional boundary conditions might be considered depending on the characteristics of the problem. Special techniques for taking into account general non-homogeneous time-dependent boundary conditions in the context of FEM have been developed in Panagiotopoulos et al. (2011), Paraskevopoulos et al. (2010).

On junctions where two rods, as shown in Fig. 9.3, of different properties (i.e., cross-section, mass density or elasticity modulus), join together, we impose the balance of force and continuity of the kinematic field (displacement, velocity). We mention here that while wave equation for a rod predicts no distortion in the propagated wave, dispersive effects in rods may arise, when considering discontinuous cross sections.

9.2 Numerical Implementation

Several numerical methods have been developed for computing the solution of wave propagation problems in structures. We just namely mention Galerkin FEM (Le Guennec and Savin 2011), Spectral Element Method (Doyle 1989; Gopalakrishnan et al. 2008), as well as, other mixed formulations of FEM (Bécache et al. 2002). Here we have adopted conventional FEM, first because engineering community is much more familiar to that and also because of its versatility regarding spatial and time discretization (Cook et al. 2001).

More specifically in our numerics we use stiffness and mass matrices for Timoshenko beam as given in Przemieniecki (1968), together with the familiar Newmark time integration algorithm (Bathe 2006) while other energy conserving algorithms (Simo and Tarnow 1992) have also been tested.

As it will be made clear later, a crucial quantity in our framework is the rate of mechanical energy. That energy for an one dimensional domain, in which we have axial and transverse deformation, as well as, rotation is given by,

$$\mathcal{E}(t) = \frac{1}{2} \underbrace{\int_0^L \rho A \left(\frac{\partial u}{\partial t}\right)^2 + \rho A \left(\frac{\partial v}{\partial t}\right)^2 + \rho I \left(\frac{\partial \psi}{\partial t}\right)^2 dx}_{\text{kinetic energy}}$$

$$+ \frac{1}{2} \underbrace{\int_0^L EA \left(\frac{\partial u}{\partial x}\right)^2 + \kappa GA \left(\frac{\partial v}{\partial x} - \psi\right)^2 + EI \left(\frac{\partial \psi}{\partial x}\right)^2 dx}_{\text{potential elastic energy}}. \tag{9.7}$$

We will further refer to the one half of the integrand quantity as the density of the total energy rate, denoted by $e(x,t)$.

9.2.1 Discretized Equations of Motion

After spatial discretization using the FEM, system of equations (9.1) and (9.5) may be written in an algebraic (matrix) form as

$$\mathbf{M\ddot{u}} + \mathbf{Ku} = \mathbf{f}, \tag{9.8}$$

where $\mathbf{M}$ and $\mathbf{K}$ are the mass and stiffness $N \times N$ matrices, $\mathbf{u}$ is the $N \times 1$ vector of unknown degrees of freedom (dofs) of the system (axial and transverse displacements (u, v) as well as the slope ψ) while dotted variables denote derivatives with respect to time. Furthermore, $\mathbf{f}$ is the $N \times 1$ vector of external loading. Up to now we have not considered any damping, however all structural systems involve some damping and for this case the respective system of equations would be,

$$\mathbf{M\ddot{u}} + \mathbf{C\dot{u}} + \mathbf{Ku} = \mathbf{f}, \tag{9.9}$$

with $\mathbf{C}$ a damping matrix. Accurate assumption on damping of a structural system is not always an easy task, however a common hypothesis is that of Rayleigh's damping (Clough and Penzien 1993) given as $\mathbf{C} = a_m\mathbf{M} + a_k\mathbf{K}$.

Equations (9.8) and (9.9) are known as the semi-discrete equations of motion since they are only discretized with respect to the spatial variables (and not the time). In principle, any standard procedure for the solution of differential equations with constant coefficients, could serve in order to deal with these systems of equations. However, it is more convenient to use a direct integration method, where Eq. (9.8) (or (9.9)) are integrated using a numerical step-by-step procedure. Two are the main components of such direct integration techniques, first, instead of trying to satisfy (9.9) at any time t, it is aimed to be satisfied only at discrete times Δt apart. The second idea is that a variation of the kinematic field, within each time interval Δt, is assumed. For an extensive presentation the interested reader is referred to Belytschko and Hughes (2014).

For linear systems such as (9.8) and (9.9), where the principle of superposition holds, the solution can be obtained by Fourier transform (frequency domain) procedures, as well as, by applying convolution integral (time domain) methods. In what follows, in Sects. 9.2.2 and 9.2.3, we mainly reproduce well known material found in the literature (Clough and Penzien 1993). In subsequent Sects. 9.3 and 9.4, we clarify the reason for which we need such an approach. In such a case, one would need the unit-impulse transfer functions that actually are the Green's function analogue for the discretized systems. Here we numerically obtain these functions by utilizing the FEM. A thorough study on the relation of Green's functions and FEM for the case of quasi-static conditions may be found in recent literature (Hartmann 2013).

9.2.2 Time Domain Formulation

Assuming that the system is subjected to a unit-impulse loading in the jth dof, while no other loads are applied, the force vector $\mathbf{f}(t)$ consists only of zero components except for the jth term which is expressed by $f_j(t) = \delta(t)$. Solving (9.8) for this specific loading, the ith component in the resulting displacement vector will then be the free-vibration response in that dof caused by a unit-impulse in coordinate j. Therefore by definition this ith component motion is a unit-impulse transfer function, which will be denoted herein by $h_{ij}(t)$.

If the corresponding loading in the jth dof was a general time varying load $f_j(t)$ rather than a unit-impulse loading, the dynamic response for the ith dof could be obtained by superposing the effects of a succession of impulses in the manner of the Duhamel's integral, assuming zero initial conditions. The generalized expression for the response of the ith dof to the load at j is the convolution integral, as follows:

$$u_{ij}(t) = \int_0^t f_j(\tau) h_{ij}(t - \tau) \, d\tau, \qquad i = 1, 2, \dots, N \qquad (9.10)$$

and the total response for the ith dof produced by a general loading involving all components of the load vector $f(t)$ is obtained by summing the contributions from all load components:

$$u_i(t) = \sum_{j=1}^N \left[\int_0^t f_j(\tau) h_{ij}(t - \tau) \, d\tau \right], \qquad i = 1, 2, \dots, N. \qquad (9.11)$$

9.2.3 Frequency Domain Formulation

The frequency-domain analysis is similar to the time-domain procedure in that it involves superposition of the effects for dof i of a unit load applied to the dof j, however, in this case both the load and the response are harmonic. Hence, the load, has the form $f_j(t) = \exp i\omega t$, while the corresponding steady-state response for ith dof will be $\hat{h}_{ij}(\omega) \exp(i\omega t)$ in which $\hat{h}_{ij}(\omega)$ is defined as the complex frequency response transfer function.

If the loading corresponding to jth dof was a general time varying load $f_j(t)$ rather than time-harmonic, the forced vibration response of the ith dof could be obtained by superposing the effects of all the harmonics contained in $f_j(t)$. For this purpose the time domain expression of the loading is Fourier transformed to obtain

$$\widehat{f}_j(\omega) = \int_{-\infty}^{\infty} f_j(t) \exp(-i\omega t) \, dt \qquad (9.12)$$

and then by inverse Fourier transformation the responses to all of these harmonics are combined to obtain the total forced vibration response for the ith dof, as follows (assuming zero initial conditions):

$$u_{ij}(t) = \frac{1}{2\pi} \int_{-\infty}^{\infty} \widehat{h}_{ij}(\omega)\widehat{f}_j(\omega) \exp(i\omega t) \, d\omega. \qquad (9.13)$$

Finally, the total response for ith dof produced by a general loading involving all components of the load vector $\mathbf{f}$ could be obtained by superposing the contributions from all the load components:

$$u_i(t) = \frac{1}{2\pi} \sum_{j=1}^{N} \left[\int_{-\infty}^{\infty} \widehat{h}_{ij}(\omega)\widehat{f}_j(\omega) \exp(i\omega t) \, d\omega \right], \qquad i = 1, 2, \dots, N. \qquad (9.14)$$

Equations (9.11) and (9.14) constitute general solutions to the coupled equations of motion, assuming zero initial conditions. Their successful implementation depends on being able to generate the transfer functions $h_{ij}(t)$ and $\widehat{h}_{ij}(\omega)$ efficiently. While this is not practical for the time-domain functions, efficient procedures for implementing the frequency-domain formulation can be developed.

Moreover, it can also be shown that any unit impulse response transfer function $h_{ij}(t)$ and the corresponding complex frequency response transfer function $\widehat{h}_{ij}(\omega)$ are Fourier transform pairs, provided some damping is present in the system in order for the inverse transformation of Eq. (9.13) to exist.

9.3 Time Reversal Process

Assuming a medium Ω under given boundary and initial conditions, let a sub-domain of it Ω_s, where excitation is applied and a second one Ω_r, where sensors record response in time, intersection $\Omega_s \cap \Omega_r$ might be non-empty. Medium Ω might be heterogeneous, embedding a number of scatterers, and may be of complicated geometry.

According to the time reversal concept (Fink et al. 2000), an input signal can be reconstructed at an excitation point $x_s \in \Omega_s$ if an output signal recorded at a collection of points $x_r \in \Omega_r$ is re-emitted in the same medium after being reversed in the time domain. This process is referred to as time reversal and is based on the spatial *reciprocity* (also found as symmetry in engineering) and the time reversal *invariance* (under the transformation $t \to -t$) of linear wave equations. A review article on time reversal and its applications has been recently published (Givoli 2014), while a brief presentation and applications on structural systems may be found in Kohler et al. (2009). Our methodology is also described in Tsogka et al. (2015); Panagiotopoulos et al. (2015) while a different implementation for one-dimensional domains is considered in Guennec et al. (2013) using a transport equation model for the energy.

While TR was originally developed for undamped systems, it is believed and has been shown, that it may also be applied to systems with damping (Ammari et al. 2013). Solving (9.8) or (9.9) for the unit impulse force vector, corresponding to the source dof, we compute the solution at all dofs while keeping track of the dofs corresponding to x_r. That is the forward propagation of the procedure. By reversing the response at x_r in time and considering it as the new excitation imposed on the corresponding dofs, we define the backward step of the procedure. Time reversal refocusing states that the solution will refocus at the original source location at time $t = 0$.

Furthermore, considering defects as points of secondary sources we may define a methodology for defect or damage identification and localization in structures. In what follows we present some elementary results considering the above framework.

9.3.1 Instrumentation and Data Collection

In this work we describe the numerical implementation of time reversal in elastic media and carry out simulations in order to assess the effectiveness of this process in damage identification problems of structures. We assume that the medium Ω is one-dimensional or an assembly of one dimensional components. Response might be recorded on N_r sensors, belonging in $\Omega_r \in \Omega$, while excitation is assumed to be produced by N_s source points forming $\Omega_s \in \Omega$, where Ω the collection of the entire set of dofs that describe and represent the problem. These two subsets, Ω_r and Ω_s may be totally separated, coincide or just have an overlap.

In structural health monitoring (SHM) the forward step of TR corresponds to a physical process where the data response matrix is collected on a set of sensors located on the structure and being compatible with dofs u_r, $r = 1, \ldots, N_r$.

Remark that at a specific location in space, we have in general associated more than one dofs corresponding to the number of unknowns at this location. Therefore each column of the response matrix corresponds to the response received at all sensors and in all dofs, u_r, when a source point emits a pulse from a location compatible with a specific dof, u_s. In our case, the forward step is numerically simulated. A source located on the structure emits a pulse compatible with a specific dof, u_s, and the response is recorded at the receiver locations and corresponding dofs u_r. Considering that these pulses emulate impulses, the matrix constructed by this process is an approximation of the impulse response matrix (IRM) with components the $h_{ij}(t)$ given in Eq. (9.10) for $i = 1, \ldots, N_r$ and $j = 1, \ldots, N_s$. In the case where we try to identify and locate defects or damages in the structure, we may sometimes work with the "scattered" IRM which will contain the difference of the impulse response of the damaged minus the impulse response of the healthy configuration of the structure.

9.3.2   Source Localization

Consider a rectangular beam of finite length $L = 30$ m and square cross section of unit area, an elastic material of modulus $E = 1$Pa, Poisson's ratio $v = 0.25$ and mass density $\rho = 1$ kg/m^3, with both ends fixed (Dirichlet homogeneous boundary conditions). Wave propagation velocities are $c_0 = c_1 = 1$ m/s, while $c_2 = 0.578$ m/s, considering $\kappa = 5/6$. For the simulation we use a mesh of 1200 elements that means a size $h = 0.025$ m and a time step $dt = 0.025$ s. We assume a source located at $x_s = 0.75$ L having both axial and transverse components corresponding to u and v, and being a Ricker wavelet in time, given by:

$$f(x,t) = a\delta(x - x_s)\left(1 - 2\pi^2 s^2(t - t_0)^2\right)e^{-\pi^2 s^2(t-t_0)^2},$$

with $t_0 = 3$ s and $s = 1.5$, and the amplitude $a = 1000$. We also assume a sensor located at the central point $x_r = 0.5$ L recording the full kinematic field, that is, the group of dofs u, v and ψ in time. The total time of the experiment is $T = t_0 + 2L/c_1 = 63$ s. Numerically resolving the problem we compute the response's time history. Wave propagation on the length of the rod is depicted in Figs. 9.4, 9.5, 9.6 and 9.7. Furthermore, Fig. 9.8 shows the response in time as recorded on sensor at x_r for the full set of dofs. This latter response is then reversed in time and imposed as loading on x_r to solve the backward step. It is expected, and also confirmed by results in plots of Fig. 9.9, the refocusing of the wave at the original source point x_s for time $T - t_0 = 60$ s.

The quality of the refocusing can be improved by increasing the number of recording sensors as illustrated in the same Fig. 9.9 where ten and one hundred sensors, equally spaced on the length, have also been used. A clear improvement is obtained and we observe a decrease in the amplitude of the ghosts that are due to reflections of the propagating waves on the boundaries as the number of sensors increases. A quantitative analysis of the improvement in the case of acoustic waves was carried out in Tsogka et al. (2015) where we show that the signal to noise ratio (SNR), defined as the value at the source divided by the maximal ghost value, is linear with respect to the number of receivers.

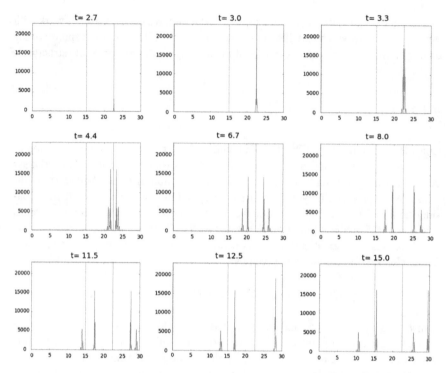

Fig. 9.4 Forward wave propagation for the rate of the total energy, the *red vertical line* shows the location of the source while the *green* the location of the sensor

9.3.3 Damage Identification

We assume here a damaged area on the rod taking place from $x_1 = 9.025$ m to $x_2 = 9.125$ m where the material has reduced modulus $E_d = 0.1$ Pa corresponding to a velocity $c_{0,d} = 0.316$ m/s. In the forward step source and receivers points, as well as, the Ricker pulse emitted, kept the same as for the previous example. We are now interested in the scattered field because of the damaged area, that is computed as the difference of the wave propagating in the damaged configuration minus the corresponding in the original reference state; The axial component motion of that scattered field is depicted in Fig. 9.10, where it can be observed that letting the pulse emitted from the source at $x_s = 22.5$ m on the initial time $t_0 = 3$ s, the wave travelling with velocity $c_0 = 1$ m/s, arrives at $x_1 = 9.125$ m at a time of about 16 s. This is the approximate time that the scattered field starts to appear. It is obvious that in this case this secondary source is of more complicated form than that of the original source shown previously, for example, in Fig. 9.5. However it is possible by using this scattered wave field and the time reversal procedure, that is, reverse in time the "recorded" scattered field and re-emit it, to construct images of the damage shown in plots of Fig. 9.11. As it can also be seen in these plots, we use axial and transverse

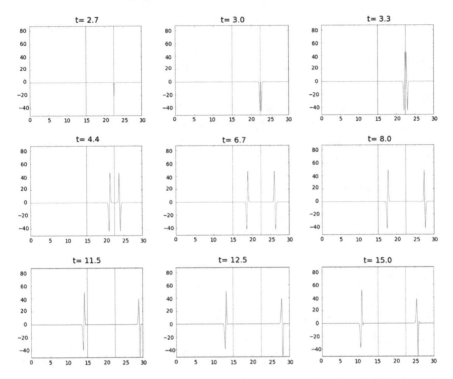

Fig. 9.5 Forward wave propagation for the axial displacement u, the *red vertical line* shows the location of the source while the *green* the location of the sensor

components to make images and locate damage, indicated by peak response, using different times, 46.3 and 37.5 s respectively, due to the fact that these two waves travel with different velocities as has been stated previously in this chapter. What else can be seen in these plots, is that by increasing the number of receiver points images are improved and the results appear to have a increased SNR. Finally, let us note that some peaks of noise that do not seem to decrease, are due to source rather than receivers and we would rather need to increase the number of source points in order to further improve the quality of image (Tsogka et al. 2015).

9.3.4   A Bridge-Like Structural Example

In this quite academic example, a bridge like frame structure is considered. Assumptions for the Timoshenko beam have been once again assumed while a modified material for both the original, as well as, the damaged configurations, has been considered. The difference, with respect to the material used before, is the mass density which in this case is $\rho = 0.1 \, \text{kg/m}^3$ and the corresponding wave propagation veloc-

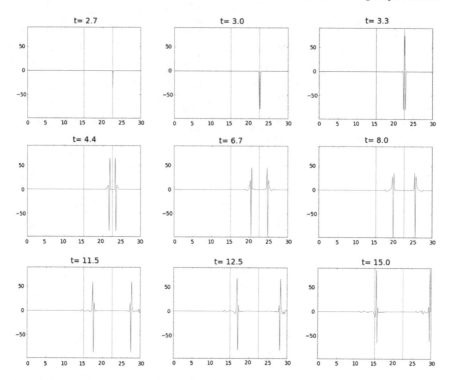

Fig. 9.6 Forward wave propagation for the transverse displacement v, the *red vertical line* shows the location of the source while the *green* the location of the sensor

ities are $c_0 = c_1 = 3.46\,\text{m/s}$, while $c_2 = 2.0\,\text{m/s}$. Once again unit cross-sections are assumed. The total maximum length of the structure is $L = 40\,\text{m}$ and its height $h = 5\,\text{m}$ while it consists of eight spans of equal length. We present only the case of damage identification, however, numerical experiments for source localization have also concluded perfect results. As regards boundary conditions we considered restrained both horizontal and vertical displacement of the furthest right and left ends of the structure. Numerical solution presented in this example is for a total number of 1560 elements of lengths 0.0833 m (vertical elements), 0.095 m (horizontal elements) and 0.1265 m (inclined elements). The total time considered is $t = 26.3875\,\text{s}$ while a time step $dt = 0.0125\,\text{s}$ has been chosen, that results a total number of 2111 time steps. Originally, it is assumed that possible positions for the sensor's placements are the junctions indicated in Fig. 9.12 as s_i, where $i \in [1, 14]$, and also it is possible for any of the three dofs to be recorded and then re-emitted. The source node, where a Ricker wavelet has been imposed, coincides with point of s_3. In Fig. 9.13 the response, numerically obtained and recorded, on sensor s_{14} for both the reference and damaged configurations, together with their difference which results the scattered field, is plotted. The scatterd field is actually the one that is reversed in

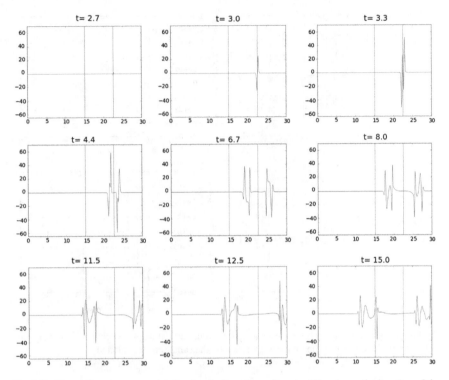

Fig. 9.7 Forward wave propagation for the slope ψ, the *red vertical line* shows the location of the source while the *green* the location of the sensor

time and re-emitted. Similar data are plotted for the central point of the damaged zone in Fig. 9.14.

We present here results for three alternative sensors' distributions, while we consider as recorded and re-emitted responses these of the two translational dofs. According to the first configuration we assume as active sensors these of s_7 and s_9, while for the second one these of s_2, s_7, s_9 and s_{12} and the final one where we assume active sensors on the whole set of possible sensors. We have also kept track for both the Euclidian norm and the energy density as possible variables to construct the image. The time evolution of these quantities are shown in the plots of Fig. 9.15. The maximum value for the Euclidian norm (left) appears at time $t = 13.0125$ s which corresponds to discrete time step $i_{nrm} = 1041$, while for the energy rate (right) at time $t = 17.0375$ s which corresponds to the discrete time step $i_{nrg} = 1363$. In order to construct an image using the energy norm, shown in Fig. 9.17, we use the i_{nrg} time step for all the three sensor configurations, while by using the Euclidian norm, shown in Fig. 9.16, both the i_{nrm} and i_{nrg} steps are used just for the case of configuration where the whole set of sensors are considered. We may also observe the fact that the reversed time $t = 26.3875 - 17.0375 = 9.35$ s approximately corresponds to the time that the scattered signal starts to be emitted at the location of the damage

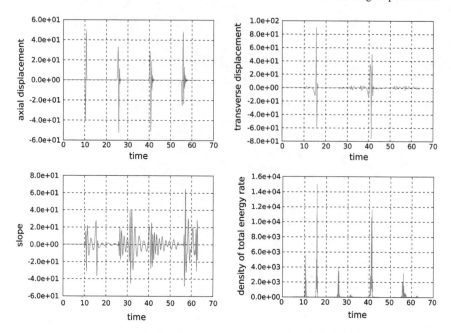

Fig. 9.8 Response recorded on a sensor at x_r for axial, vertical displacements u, v and slope ψ. Furthermore, the density of the total energy rate e on the same sensor is plotted

as a secondary source, which might be seen also in Fig. 9.14. As it can be observed in the plots of Figs. 9.16 and 9.17, the energy density appeared to be a more suitable variable for imaging in this case, since location of damage has been found very accurately. It is also observed in the case of Fig. 9.16, that improvement of imaging is achieved with increased number of sensors that record and re-emit the response.

9.4 Imaging Technique

Imaging is the discipline in science and engineering that consists of creating a representation of some medium, or structure, from recordings of waves that have propagated through (or scattered by) the medium. Imaging techniques, have a wide range of applications, e.g., optics, geophysics, medical, etc. However, it seems that it is underestimated in the field of structural engineering. Here we try to define an imaging framework capable to deal with problems of structural systems and we, briefly, give an introduction for that matter, since it is an ongoing research topic of our group. The interested reader is referred to Bleistein et al. (2001); Borcea et al. (2005) and references therein for further study.

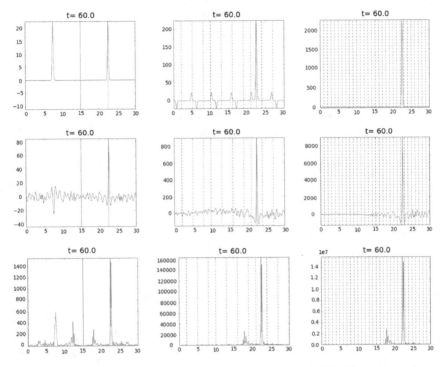

Fig. 9.9 Source refocusing after backward imposition of time reversed data. From upper to lower corresponds to u, v and rate of total energy density h, while from *left* to *right* number of receivers corresponds to 1, 10 and 100, respectively

9.4.1 Source Localization

Here we use the time reversal process in order to define an imaging functional for the discrete system that we are interested in, similar to the procedure presented in Tsogka et al. (2015) for the case of the continuous acoustic wave equation. Data at the receiver $u_r(t)$ and the solution of the backward problem at the ith dof, $u_i^{TR}(t)$, might be given, using Eq. (9.11), as:

$$u_r(t) = \int_0^t f_s(\tau) h_{rs}(t - \tau) \, d\tau, \tag{9.15}$$

$$u_i^{TR}(t) = \int_0^t u_r(T - \tau) h_{ri}(t - \tau) \, d\tau. \tag{9.16}$$

In practice we measure the data $u_r(t)$ physically or simulate it numerically. In the frequency domain, using Fourier transform we get

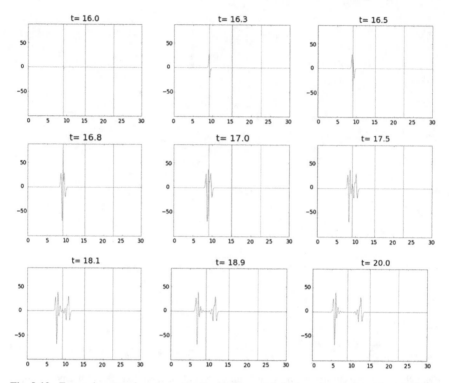

Fig. 9.10 Forward scattered wave propagation for the axial displacement u, the *red vertical line* shows the location of the source while the *green* the location of the sensor

$$u_r(t) = \frac{1}{2\pi} \int\limits_{-\infty}^{\infty} \hat{h}_{rs}(\omega)\hat{f}_s(\omega) \exp(i\omega t)\, d\omega, \quad \hat{u}_r(\omega) = \hat{h}_{rs}(\omega)\hat{f}_s(\omega), \qquad (9.17)$$

while the time reversed data in frequency domain is,

$$\hat{u}_i^{TR}(\omega) = \hat{h}_{ri}(\omega)\overline{\hat{u}_r(\omega)} = \hat{h}_{ri}(\omega)\overline{\hat{h}_{rs}(\omega)\hat{f}_s(\omega)} \qquad (9.18)$$

where overline denotes complex conjugate. Therefore, in accordance also to Eq. (9.13), the response obtained during the backward step for the ith dof can be written as,

$$u_i^{TR}(t) = \frac{1}{2\pi} \int\limits_{-\infty}^{\infty} \hat{h}_{ri}(\omega)\overline{\hat{h}_{rs}(\omega)\hat{f}_s(\omega)} \exp(i\omega t)\, d\omega. \qquad (9.19)$$

Recalling now that in time-reversal we send back the field recorded at all receivers r, see also Eq. (9.14), we get the following expression for the time reversed field at

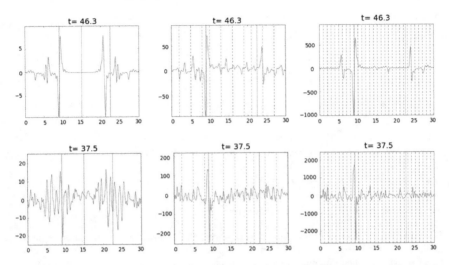

Fig. 9.11 Damage refocusing after backward imposition of time reversed data. Upper raw imaging by using data for the axial displacement, while lower for the transverse displacement. Damage area is pointed by two *vertical cyan lines*. One receiver is used for the *left column* figures while ten receivers are used to produce the medium and one hundred for the rightmost images. For the axial displacement improvement in the localisation is observed by increasing the number of receivers while a persistent ghost remains when the transverse displacement is used

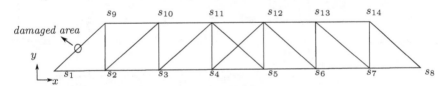

Fig. 9.12 A sketch of the bridge structure where the position of damage is depicted and also points of possible sensor placement are indicated

the ith dof,

$$u_i^{TR}(t) = \frac{1}{2\pi} \int_{-\infty}^{\infty} \sum_{r=1}^{N_r} \hat{h}_{ri}(\omega)\overline{\hat{h}_{rs}(\omega)\hat{f}_s(\omega)} \exp{(i\omega t)} \, d\omega. \tag{9.20}$$

By evaluating the time-reversed field at time $t = 0$ which is the time at which we expect refocusing at the source we obtain

$$u_i^{TR}(t = 0) = \frac{1}{2\pi} \int_{-\infty}^{\infty} \sum_{r=1}^{N_r} \hat{h}_{ri}(\omega)\overline{\hat{h}_{rs}(\omega)\hat{f}_s(\omega)} \, d\omega. \tag{9.21}$$

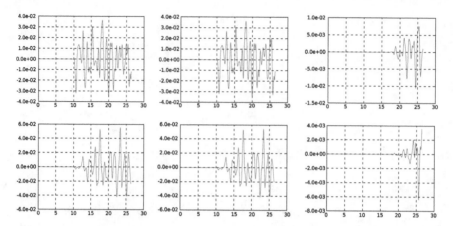

Fig. 9.13 Horizontal (*upper row*) and vertical (*lower row*) displacements recorded on sensor s_{14} during the forward process. From *left* to *right* is depicted the response on healthy (reference) structure, damaged and their difference which is the scattered field

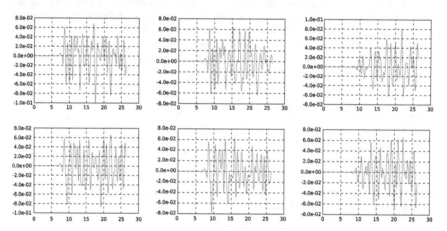

Fig. 9.14 Horizontal (*upper row*) and vertical (*lower row*) displacements' response on the damaged zone during the forward process. From *left* to *right* is depicted the response on healthy (reference) structure, damaged and their difference which is the scattered field

This motivates us to define an imaging functional as

$$I_m = \frac{1}{2\pi} \sum_{\omega} \sum_{r=1}^{N_r} \overline{\widehat{h}_{rs}(\omega)\widehat{f}_s(\omega)}\widehat{h}_{rm}(\omega) = \frac{1}{2\pi} \sum_{\omega} \sum_{r=1}^{N_r} \overline{\widehat{u}_r(\omega)}\,\widehat{h}_{rm}(\omega) \qquad (9.22)$$

which associates a value at the mth dof by back propagating the recordings (data) reversed in time, $\overline{\widehat{u}_r(\omega)}$, using all receivers and all available frequencies.

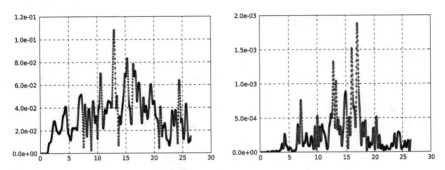

Fig. 9.15 Evolution in time of the Euclidian norm of response (*left*) and energy rate (*right*) on the damaged zone

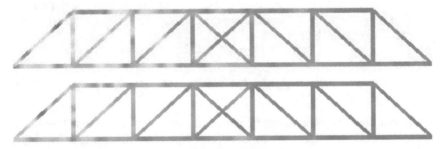

Fig. 9.16 Euclidian norm as an image of the damage location for appropriate times selected from maximum value of energy (*upper*) and euclidean (*lower*) norm evolutions. Sensors for recording considered on every s_i for $i \in [1, 14]$

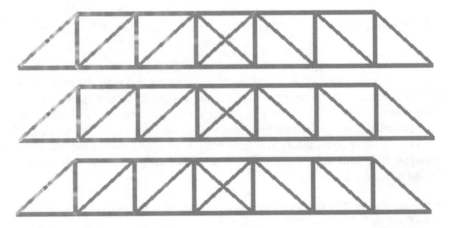

Fig. 9.17 Energy density on appropriate time, selected from maximum value of energy evolution, as an image of the damage location for the three configurations of active sensors. Number of sensor is increased from *upper* to *bottom* pictures

9.4.2 Damage Identification

Equivalently to the source localization process, in the present subsection we perform the backward step of the defect or damage localization problem in the frequency domain. For that purpose we consider the scattered field at the receivers, and based on existing models for the continuous medium (Borcea et al. 2005; Tsogka et al. 2015), we define an imaging functional, for the discrete system case, as

$$I_m = \sum_\omega \sum_{r=1}^{N_r} \overline{\hat{u}_{rs}^{\text{sc}}(\omega)} \, \hat{h}_{rm}(\omega) \hat{h}_{rs}(\omega) \tag{9.23}$$

Here $\hat{u}_{rs}^{\text{sc}}(\omega)$ is the scattered field recorded at dof r due to an excitation at dof s. If we have data for multiple excitations (sources) we superpose the images obtained for each one of them and obtain,

$$I_m = \sum_\omega \sum_{r=1}^{N_r} \sum_{s=1}^{N_s} \overline{\hat{u}_{rs}^{\text{sc}}(\omega)} \, \hat{h}_{rm}(\omega) \hat{h}_{rs}(\omega). \tag{9.24}$$

We make clear at this point that I_m refers to the mth dof of the system. One might want to construct an image using only a specific type of dof on each spatial location or we might consider images of quantities such as the energy that combine more than one dofs.

9.5 Conclusions

We presented in this chapter a methodology for localizing sources, as well as, small defects and/or damaged areas, found on frame structures that could be modeled by beam elements. Time reversal procedure and standard imaging techniques have been defined and presented. Numerical implementation for time reversal experiments have been formulated, using standard finite element approximations, by the adoption of the Timoshenko's beam theory and time integration algorithms. Some numerical examples have been solved and results have been indicatively demonstrated and proved to be very promising.

Acknowledgements This work was partially supported by the European Research Council Starting Grant Project ADAPTIVES-239959.

References

Ammari H, Bretin E, Garnier J, Wahab A (2013) Time-reversal algorithms in viscoelastic media. Eur J Appl Math 24(04):565–600

Anderson BE, Griffa M, Larmat C, Ulrich TJ, Johnson PA (2008) Time reversal. Acoust Today 4(1):5–16

Bathe K-J (2006) Finite element procedures. Klaus-Jurgen Bathe

Bécache E, Joly P, Tsogka C (2002) A new family of mixed finite elements for the linear elastodynamic problem. SIAM J Numer Anal 39:2109–2132

Belytschko T, Hughes TJ (2014) Computational methods for transient analysis. Comput Methods Mech 1

Bleistein N, Cohen J, John W (2001) Mathematics of multidimensional seismic imaging, migration, and inversion. Springer Science+Business Media, New York

Borcea L, Papanicolaou G, Tsogka C (2005) Interferometric array imaging in clutter. Inverse Probl 21(4):1419

Clough RW, Penzien J (1993) Dynamics of structures. McGraw-Hill, Singapore

Cook RD, Malkus DS, Plesha ME, Witt RJ (2001) Concepts and application of finite element analysis, 4th edn. Wiley, United States

Cowper GR (1966) The shear coefficient in Timoshenko's beam theory. J Appl Mech 33:335–340

Doyle JF (1989) Wave propagation in structures: an FFT-based spectral analysis methodology. Springer, New York

Fink M, Cassereau D, Derode A, Prada C, Roux P, Tanter M, Thomas J-L, Wu F (2000) Time-reversed acoustics. Rep Prog Phys 63(12):1933

Fink M, Prada C (2001) Acoustic time-reversal mirrors. Inverse Probl 17(1):R1

Fung YC (1965) Foundations of solids mechanics. Prentice-Hall, Englewood Cliffs, New Jersey

Givoli D (2014) Time reversal as a computational tool in acoustics and elastodynamics. J Comput Acoust 22(03)

Gopalakrishnan S, Chakraborty A, Mahapatra DR (2008) Spectral finite element method: wave propagation, diagnostics and control in anisotropic and inhomogeneous structures. Springer, London

Graff KF (1975) Wave motion in elastic solids. Dover publications, New York

Guennec YL, Savin E, Clouteau D (2013) A time-reversal process for beam trusses subjected to impulse loads. J Phys Conf Ser 464(012001)

Hartmann F (2013) Green's functions and finite elements. Springer, Berlin

Kohler MD, Heaton TH, Heckman V (2009) A time-reversed reciprocal method for detecting high-frequency events in civil structures with accelerometer arrays. In: Proceedings of the 5th international workshop on advanced smart materials and smart structures technology

Le Guennec Y, Savin É (2011) A transport model and numerical simulation of the high-frequency dynamics of three-dimensional beam trusses. J Acoust Soc Am 130(6):3706–3722

Panagiotopoulos CG, Paraskevopoulos EA, Manolis GD (2011) Critical assessment of penalty-type methods for imposition of time-dependent boundary conditions in fem formulations for elastodynamics. In: Computational methods in earthquake engineering. Springer, pp 357–375

Panagiotopoulos CG, Petromichelakis Y, Tsogka C (2015) Time reversal in elastodynamics with application to structural health monitoring. In: Proceedings of the 5th international conference on computational methods in structural dynamics and earthquake engineering

Paraskevopoulos E, Panagiotopoulos C, Manolis G (2010) Imposition of time-dependent boundary conditions in fem formulations for elastodynamics: critical assessment of penalty-type methods. Comput Mech 45:157–166

Prada C, Thomas J-L, Fink M (1995) The iterative time reversal process: analysis of the convergence. J Acoust Soc Am 97(1):62–71

Przemieniecki J (1968) Theory of matrix structural analysis. Dover publications, Inc., New York

Simo J, Tarnow N (1992) The discrete energy-momentum method. Conserving algorithms for nonlinear elastodynamics. J Appl Math Phys 43:757–792

Timoshenko SP (1921) On the correction for shear of the differential equation for transverse vibrations of bars of uniform cross-section. Phil Mag 41:744–746

Timoshenko SP (1922) On the transverse vibrations of bars of uniform cross-section. Phil Mag
43:125–131

Tsogka C, Petromichelakis Y, Panagiotopoulos CG (2015) Influence of the boundaries in imaging for damage localization in 1D domains. In: Proceedings of the 8th GRACM international
congress on computational mechanics

Yavuz ME, Teixeira FL (2009) Ultrawideband microwave sensing and imaging using time-reversal
techniques: a review. Remote Sens 1(3):466–495

Chapter 10
Decentralized Infrastructure Health Monitoring Using Embedded Computing in Wireless Sensor Networks

Kosmas Dragos and Kay Smarsly

Abstract Due to the significant risk posed to public safety by infrastructure ageing and deterioration, infrastructure health monitoring has been drawing increasing research interest in recent years. In the field of infrastructure health monitoring, wireless sensor networks have become particularly popular, because of their cost efficiency, flexibility, and reduced installation time as compared to conventional wired systems. Using embedded computing, it is possible to process the collected data directly on the wireless sensor nodes, thus reducing the wireless communication and the power consumption. In this paper, a methodology to fully decentralize the condition assessment process is presented. The paper showcases the development of embedded algorithms and numerical models to be embedded into the wireless sensor nodes, i.e. the monitored structure is divided into substructures where each sensor node is responsible for monitoring one substructure. The proposed methodology is validated through simulations on a numerical model of a four-story shear frame structure. The objective of the simulations is to test the performance of the algorithms and the quality of the numerical models embedded into the wireless sensor nodes with respect to decentralized condition assessment, taking into consideration the effects of external factors, such as ambient noise, that usually interfere with measured data.

Keywords Infrastructure monitoring · Wireless sensor networks · Decentralized systems · Structural dynamics

10.1 Introduction

The risk to public safety associated with the deteriorating condition of civil infrastructure has fueled research on condition assessment of structures and has led to the development of the field of structural health monitoring (SHM). In con-

K. Dragos · K. Smarsly (✉)
Chair of Computing in Civil Engineering, Bauhaus University Weimar, 99423
Weimar, Germany
e-mail: kay.smarsly@uni-weimar.de

© Springer International Publishing AG 2017
A.G. Sextos and G.D. Manolis (eds.), *Dynamic Response of Infrastructure to Environmentally Induced Loads*, Lecture Notes in Civil Engineering 2,
DOI 10.1007/978-3-319-56136-3_10

ventional SHM systems, wired sensors are deployed for collecting data to be used for the condition assessment of structures. However, the installation of wired SHM systems has been proven labor-intensive and costly due to the need for expensive coaxial cables, particularly in large structures. It has been reported that the cost for the installation of a wired SHM system could reach the amount of $5,000 per sensing channel (Celebi 2002). As a consequence, the civil engineering community has been pursuing more efficient alternatives, such as wireless sensing technologies.

The use of wireless sensor networks (WSNs) for SHM has been gaining increasing attention in recent years. Wireless communication among the sensor nodes eliminates the need for cable connections, thus notably reducing the cost and time needed for installation. Furthermore, utilizing the advantage of collocated sensing modules with processing units on-board the wireless sensor nodes, data processing prior to the wireless transmission is possible. On-board processing of the collected data leads to a reduced amount of data to be wirelessly communicated, entailing a significant reduction in power consumption, which is still a major constraint in WSNs. Other drawbacks of WSNs are related to the reliability of wireless transmission and to the synchronization of data, which can be solved by implementing adequate embedded computing capabilities.

The embedded computing capabilities of wireless sensor nodes have already been utilized from the early stages of WSN applications in SHM. For example, Lynch et al. (2004) proposed the use of an autoregressive model with exogenous inputs (AR-ARX) for damage detection. Using the same sensor node prototype, Wang et al. (2007) introduced multi-threaded embedded software for the execution of simultaneous tasks on the sensor nodes. In the field of system identification, Zimmerman et al. (2008) proposed embedded algorithms for the execution of output-only system identification methods, while Cho et al. (2008) presented the wireless tension force estimation system for cable forces in cable-stayed bridges. A simulated annealing algorithm for model updating was presented by Zimmerman and Lynch (2007). Furthermore, Lei et al. (2010) demonstrated a significant reduction in power consumption as a result of the incorporation of data processing algorithms into the sensor nodes. In structural control applications, the use of a linear quadratic regulation algorithm was proposed by Wang et al. (2006) and Kane et al. (2014). Finally, in distributed networking approaches, Rice et al. (2010) presented the "Illinois structural health monitoring project" tool suite, which offers a variety of services related to data collection, data processing, and communication reliability. The use of neural networks for autonomous fault detection, making use of the inherent redundancy in sensor outputs, was proposed by Smarsly and Law (2013b). The same group presented a migration-based approach (Smarsly and Law 2013a), where powerful software agents are automatically assembled in real time to migrate to the sensor nodes in order to analyze potential anomalies on demand in a resource-efficient manner.

In summary, the aforementioned embedded computing approaches cover a broad range of SHM tasks; however, for a fully decentralized SHM system, the intelligence of the wireless sensor network and the ability of smart sensor nodes to autonomously perform SHM tasks needs to be enhanced. In this paper, a methodology for decentralized condition assessment of civil infrastructure is presented. Exploiting the processing power of wireless sensor nodes, the embedment of a decentralized numerical model comprising coupled "partial" numerical models, i.e. sub-models of the overall model, is proposed in order to enhance the ability of the sensor nodes to perceive the physical characteristics of the monitored structure. The proposed methodology consists of two stages. Similar to conventional SHM approaches, the first stage is the system identification performed to establish the current state of the structure in the form of a numerical model that serves as reference ("model updating"). The second stage is the assessment of the structural condition by analyzing whether the newly collected acceleration response data fits the structural parameters of the model obtained in the first stage ("condition assessment").

In the first part of the paper, the theoretical background of the proposed methodology and the techniques used for system identification are presented. The merits of using embedded models are explained and the steps of the methodology are outlined. The second part of the paper covers the implementation and validation of the proposed methodology. The architecture of the wireless sensor network developed in this study is described and the embedded software is presented. The methodology is validated through simulations on a four-story shear frame structure. Finally, the test results are discussed and an outlook on potential future extensions is given.

10.2   A Methodology Enabling Decentralized Condition Assessment of Civil Infrastructure

The methodology for decentralized condition assessment of civil infrastructure comprises two stages. In the first "model updating" stage, the wireless sensor network establishes an "initial" decentralized numerical model of the monitored structure representing the initial structural state in terms of stiffness and damping parameters. In the second stage, the "condition assessment" stage, the network collects acceleration response data from an unknown structural state. The wireless sensor nodes of the network check whether the parameters of the initial model fit the newly collected data. Potential deviations exceeding a predefined threshold could indicate damage. A detailed description of both stages is given in the following subsections.

10.2.1 Model Updating

The objective of the model updating stage is to derive a numerical model, distributed to different sensor nodes, that represents the initial state of the monitored structure. The parameters of the initial model are automatically calculated by the sensor nodes by employing an embedded system identification method. In the following subsections, system identification methods are briefly discussed, the proposed method is presented, and the steps of the model updating stage are described.

System identification

System identification methods have been extensively used in SHM. Among these methods, vibration-based methods, which make use of acceleration response data, are particularly popular in civil engineering. Conventional vibration-based methods are applied through force vibration testing (FVT), where structures are artificially excited and both the input and the output of the test are known. Alternatively, ambient vibration testing (AVT) methods are based on natural (ambient) excitation with unknown input, by employing output-only methods and assuming that the input is zero-mean Gaussian white noise. Given the difficulties in exciting large civil engineering structures, there is a tendency towards a prevalence of AVT methods over FVT methods (Cunha et al. 2005).

For extracting structural properties, data processing in system identification is performed either in the frequency domain or in the time domain. The most common method used in the frequency domain is the "frequency domain decomposition" (FDD) proposed by Brincker et al. (2000). The FDD method obtains estimates of the mode shape vectors through singular value decomposition of the spectral density matrix of the output, which is easily calculated in case of zero-mean Gaussian white noise excitation. An example of time domain methods is stochastic subspace identification (SSI), which deals with the extraction of mode shapes by fitting a "system matrix" directly to the response data (Peeters and De Roeck 1999). There are several other methods to perform system identification both in the frequency domain and in the time domain, but further description of those methods falls beyond the scope of this chapter.

A methodology for system identification

The model updating stage of the condition assessment methodology is associated with the establishment of the initial numerical model. In order to achieve the highest possible degree of decentralization in the condition assessment process autonomously conducted by wireless sensor nodes, the monitored structure is divided into substructures. Each wireless sensor node is responsible for assessing the condition of a substructure. It should be noted that in this study the condition assessment process is decentralized; hence each sensor node must be able to perceive the behavior of the substructure it is responsible for. A partial numerical model of the structure is therefore necessary for each substructure.

Following the principles of the finite element method (FEM), each substructure is discretized according to the number of sensing units to be attached to each substructure. The estimation of the stiffness parameters is performed by solving the dynamic equilibrium equations using response data from free vibrations. In general form, the dynamic equilibrium equations of a structural system with N degrees of freedom (DOFs) is given in Eq. 10.1.

$$\mathbf{M}_{N \times N} \cdot \ddot{\mathbf{u}}(t)_N + \mathbf{C}_{N \times N} \cdot \dot{\mathbf{u}}(t)_N + \mathbf{K}_{N \times N} \cdot \mathbf{u}(t)_N = \mathbf{F}(t)_N \qquad (10.1)$$

In Eq. 10.1, $\mathbf{M}$, $\mathbf{C}$, and $\mathbf{K}$ are the mass matrix, the damping matrix, and the stiffness matrix, respectively, while $\ddot{\mathbf{u}}(t)$, $\dot{\mathbf{u}}(t)$, $\mathbf{u}(t)$ are the acceleration vector, the velocity vector and the displacement vector, respectively. $\mathbf{F}(t)$ is the external force vector.

Using the fast Fourier transform (FFT), Eq. 10.1 can be transformed into the frequency domain (Eq. 10.2). Instead of the acceleration, velocity and displacement vectors, the respective Fourier amplitudes of these vectors can be used at the frequencies ω that correspond to modes of vibration.

$$\mathbf{M}_{N \times N} \cdot \ddot{\mathbf{u}}(\omega)_N + \mathbf{C}_{N \times N} \cdot \dot{\mathbf{u}}(\omega)_N + \mathbf{K}_{N \times N} \cdot \mathbf{u}(\omega)_N = \mathbf{F}(\omega)_N \qquad (10.2)$$

First, acceleration response data is collected under free vibration. Second, the obtained acceleration data is integrated, using numerical integration methods, and the corresponding velocities and displacements are derived by each sensor node. Third, the accelerations, velocities and displacements are transformed into the frequency domain using the FFT, and the corresponding frequency spectra are derived. Using Eq. 10.2, the dynamic equilibrium equations of an arbitrary substructure with N degrees of freedom under free vibration are given in Eq. 10.3.

$$\mathbf{M}_{N \times N} \cdot \ddot{\mathbf{u}}_N + \mathbf{C}_{N \times (R+N+S)} \cdot \left\{ \dot{\mathbf{u}}_R \quad \dot{\mathbf{u}}_N \quad \dot{\mathbf{u}}_S \right\}^T + \mathbf{K}_{N \times (R+N+S)} \cdot \left\{ \mathbf{u}_R \quad \mathbf{u}_N \quad \mathbf{u}_S \right\}^T = \mathbf{0}_N$$

$$(10.3)$$

As can be seen from Eq. 10.3, the absence of constants leads to trivial solutions. To avoid this problem, reasonable assumptions are made for one of the terms on the left hand side of Eq. 10.3. In this study, assumptions are made about the mass matrix, and the stiffness and damping matrices are estimated accordingly.

Assuming that the division of the structure into substructures is performed in such a way that each substructure has two interfaces, each connecting the substructure with one neighboring substructure, R is used to denote the DOFs of the first interface and S is used to denote the DOFs of the second interface. Since Eq. 10.3 describes free vibration, the external force is zero.

From Eq. 10.3, a partial "hybrid" model corresponding to the substructure under consideration is generated on each wireless sensor node. It is evident that solving Eq. 10.3 on each substructure is only possible if the frequency spectral peaks of velocities and displacements of the DOFs at the interface with neighboring substructures are communicated between the wireless sensor nodes. Wireless communication is ensured through reliable links established between sensor nodes located in neighboring substructures (Fig. 10.1). An additional communication link

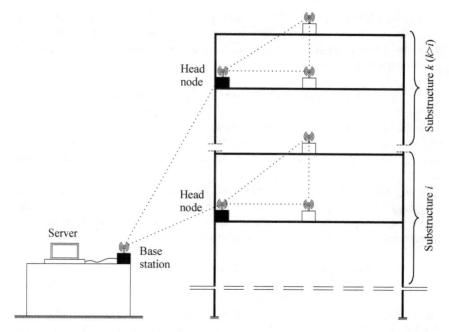

Fig. 10.1 Network architecture

is established between the server and one of the sensor nodes, designated as "head node". Depending on the size of the structure, substructures can be clustered into groups with one head node each. A head node receives commands from the server and tasks the rest of the sensor nodes of the group accordingly; hence, the exchange of velocity and displacement spectral peaks is initiated by the head nodes. The overall network architecture of the SHM system proposed in this study is illustrated in Fig. 10.1.

The unknown parameters of Eq. 10.3 are the elements of matrices $\mathbf{C}$ and $\mathbf{K}$. Each row of matrices $\mathbf{C}$ and $\mathbf{K}$ has a total of $R + N + S$ unknowns such that the required order of the system of dynamic equilibrium equations is $O = 2\,[R + N + S]$. Thus, an adequate number $n\,(n \geq O)$ of modal peaks in the frequency spectra of acceleration, velocity and displacement is selected for solving the system of equations.

Sequence of the model updating stage

The sequence of the model updating stage implemented into the wireless sensor network is described by the following steps.

A. *Initializing the model updating algorithm.* The model data for each substructure is selected and loaded to the server.
B. *Switching head nodes to "model updating mode".* A beacon signal indicating the desired mode of operation is sent from the server to the head nodes of the WSN in order to set the wireless sensor nodes to model updating mode.

C. *Switching sensor nodes to "model updating mode".* The head nodes send signals to the sensor nodes to set the nodes to model updating mode.

D. *Receiving "acknowledgement" signals.* The head nodes receive acknowledgement signals from all sensor nodes and, subsequently, send acknowledgement signals to the server.

E. *Loading model data into the network.* Model data is packaged for each substructure and sent to the corresponding sensor nodes.

F. *Receiving "acknowledgement" signals for data loading.* Each sensor node sends an acknowledgement signal to the server once the model data is loaded and switches to standby mode waiting for the excitation of the structure.

G. *Acceleration sampling.* Once a predefined acceleration threshold is exceeded, the sensor nodes start to sample acceleration data.

H. *Calculating velocities and displacements.* As soon as a predefined number of acceleration response data is collected, acceleration sampling stops. Then, velocities and displacements are calculated using a time integration method, here the Newmark-β algorithm (Newmark 1959).

I. *Transforming data to the frequency domain.* The acceleration, velocity and displacement data is transformed into the frequency domain using the FFT algorithm.

J. *Transmitting data between neighboring sensor nodes.* A predefined number of velocity and displacement frequency spectral peaks, according to the procedure described above, are exchanged between neighboring sensor nodes in order to set up the system of equations.

K. *Solving the system of equations.* The system of equations is solved by each wireless sensor node to calculate the estimates of the damping and stiffness parameters.

10.2.2 Condition Assessment

The objective of the condition assessment stage is to assess the current, i.e. unknown, condition of the structure using the model (and the embedded partial models, respectively) derived from the model updating stage as a reference. During the condition assessment stage of the wireless SHM system, a new set of acceleration response data under free vibration is collected by each sensor node, and the corresponding velocity and displacement vectors are calculated as in the model updating stage. The newly collected acceleration response data as well as the calculated velocity and displacement data are transformed into the frequency domain and used to apply the dynamic equilibrium equations with the stiffness and damping parameters of the initial model. Small errors, i.e. deviations from equilibrium in the condition assessment stage, are expected due to numerical instability and noise interference. Errors exceeding a predefined threshold could indicate damage.

The sequence of the condition assessment stage as performed by the wireless sensor network is described by the following steps.

A. *Initializing the condition assessment algorithm.* The server is started.
B. *Switching head nodes to "condition assessment mode".* A beacon signal is sent from the server to the head nodes in order to set the wireless sensor nodes to condition assessment mode.
C. *Switching sensor nodes to "condition assessment mode".* The head nodes send signals to the sensor nodes to set them to condition assessment mode.
D. *Receiving "acknowledgement" signals.* The head nodes receive acknowledgement signals from all sensor nodes and, subsequently, send acknowledgement signals to the server.
E. *Switching sensor nodes to standby mode.* The sensor nodes go to standby mode waiting to start sampling.
F. *Acceleration sampling.* Once a predefined acceleration threshold is exceeded, the sensor nodes start sampling acceleration data.
G. *Calculating velocities and displacements.* As soon as a predefined number of acceleration response data is collected, acceleration sampling stops. Then, velocities and displacements are calculated using the Newmark-β time integration method.
H. *Transforming data to the frequency domain.* The acceleration, velocity and displacement data is transformed into the frequency domain using the FFT algorithm.
I. *Transmitting data between neighboring sensor nodes.* A predefined number of velocity and displacement frequency spectral peaks, corresponding to the modes of vibration of the initial model, are exchanged between neighboring sensor nodes.
J. *Comparing the results with model updating stage.* The Fourier amplitudes of the newly collected acceleration response data and of the calculated velocity and displacement data are used to apply the dynamic equilibrium equations with the stiffness and damping parameters of the initial model derived from the model updating stage.
K. *Sending damage detection signal.* A residual, i.e. small deviation from equilibrium, is expected in the result of the dynamic equilibrium equations of step J due to noise interference and the approximations of the numerical integration algorithm. If the residual exceeds a predefined threshold, a damage detection signal is sent from the sensor node to the server.

10.3   Validation of the Decentralized Condition Assessment Methodology

The methodology proposed for decentralized condition assessment of civil infrastructure is validated through simulations conducted on a numerical model of a four-story shear frame structure. In this section, the implementation of the methodology into a wireless SHM system is presented, and the simulations are described.

10.3.1   Implementation

Following the steps of the methodology presented in the previous section, embedded software is designed in order to implement the methodology into a wireless SHM system. The software, written in Java programming language, is launched at the server with the initialization of a "host application" software, i.e. Java classes that perform the tasks of the server. Tasks executed by the wireless sensor nodes are handled by another set of Java classes embedded into the nodes, termed "on-board application" software. When starting the wireless SHM system, peer-to-peer communication links are established between neighboring sensor nodes. The mode of operation, "model updating" or "condition assessment", can be chosen by the user. Once the mode of operation is selected, a beacon signal, ensuring time synchronization of the SHM system, is sent from the server to the head nodes of the SHM system. The beacon signal is successively forwarded by the head node to the neighboring nodes until the beacon signal has reached every sensor node of the network. An acknowledgement signal is sent from the outermost node to its neighboring node and forwarded successively until the acknowledgement signal has reached the server, being notified that all sensor nodes are set to the selected mode.

If "model updating" mode is selected, data related to the partial model of each substructure is automatically loaded to the corresponding sensor node. The data is loaded over the air, i.e. through the radio communication links established between the server and the sensor nodes. The on-board applications set the nodes to standby mode and sampling starts as soon as a predefined acceleration threshold is exceeded. Once a predefined number of acceleration data is collected, sampling stops. The on-board applications proceed with the calculation of velocity and displacement data, using the Newmark-β algorithm, and transform the acceleration, velocity, and displacement data into the frequency domain. Then, velocity and displacement frequency spectral peaks are exchanged between neighboring nodes to form the system of dynamic equilibrium equations of each substructure on the respective sensor node. Finally, the system of equations is solved and the stiffness and damping parameters of the initial model are calculated.

If "condition assessment" mode is selected, the steps of acceleration sampling, numerical integration for obtaining velocities and displacements, the FFT, and the exchange of data are the same as in the model updating mode. Finally, the dynamic equilibrium equations are applied at the frequency spectral peaks corresponding to the modes of vibration of the initial model, using the stiffness and damping parameters of the initial model. If the deviations from equilibrium exceed a pre-defined threshold, a "damage detection" signal is sent to the server, indicating the substructure associated with the detected damage.

10.3.2   Validation of the Methodology

The performance of both stages of the methodology, model updating and condition assessment, depends on the accuracy of the collected acceleration response data and on the stability of the numerical integration algorithm. Hence, it is evident that the interference by external factors, such as noise present in the measurements, could be detrimental to both the quality of the generated parameters of the initial model and to the ability of the system to detect damage. As field measurements are usually contaminated with ambient noise, it is important to consider the effect of noise when validating the performance of the SHM system through simulations.

In this section, the validation of the algorithms of the methodology is presented. A simulation-based validation is performed using a finite element model of a four-story shear frame structure with known structural parameters. First, a brief description of the sensor node platform used in the proposed SHM system is given. Second, the model updating stage and the overall SHM system are described, illustrating the generation of an initial numerical model of the shear frame structure by the sensor nodes. Finally, damage on the shear frame structure is simulated to validate the condition assessment capabilities of the SHM system.

Wireless sensor node platform

The wireless sensor nodes used for the implementation of the proposed method-ology are the Oracle SunSPOTs (Small Programmable Object Technology, Oracle Corp. 2007), shown in Fig. 10.2. The hardware platform has been proven a reliable and efficient means for rapid prototyping of embedded monitoring applications in different engineering disciplines, such as structural health monitoring (Dragos and Smarsly 2015; Smarsly 2014; Chowdhury et al. 2014; Smarsly and Petryna 2014), infrastructure monitoring (Law et al. 2014; Smarsly et al. 2011), landslide moni-toring (Georgieva et al. 2012; Smarsly et al. 2012, 2014), and ecosystem moni-toring (Smarsly 2013; Smarsly and Law 2012). The wireless sensor nodes feature an ARM 920T microcontroller with a 32-bit bus size running at 400 MHz, 1 MB flash memory, and 512 kB RAM, while the operating system is the Java pro-grammable Squawk Virtual Machine. An 8-bit MMA7455L accelerometer is integrated into the sensor node platform, which can be set to sample at a maximum range of ± 2 g, ± 6 g, or ± 8 g.

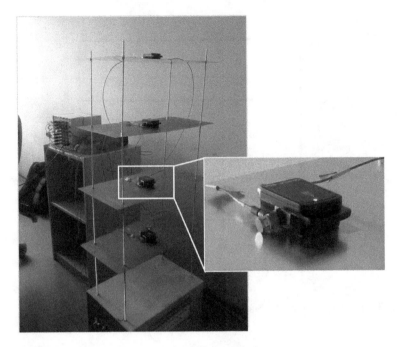

Fig. 10.2 Wireless sensor nodes mounted on the shear frame structure

Model updating stage

The four-story shear frame structure used for the simulations is illustrated in Fig. 10.3, and the structural parameters of each story are summarized in Table 10.1. The following simplifications are assumed.

- Mass is concentrated at the mid-span of each story
- A diaphragm constraint is assumed on each story, such that $u_{x,l} = u_{x,r}$, $u_{x,l}$ and $u_{x,r}$ being the horizontal displacement on the left end and on the right end, respectively, of each story
- "Shear frame structure" function is assumed, i.e. joint rotations are ignored
- Vertical displacements are considered negligible
- Modulus of elasticity is set to $E = 2 \times 10^7$ kN/m² and Poisson's ratio is set to $\nu = 0.25$
- Frame elements are discretized using the Timoshenko beam assumption (Timoshenko 1921)

Damping is considered proportional to mass and stiffness (Rayleigh 1877). A proportionality coefficient of $\alpha_1 = 0.5$ is selected for the mass matrix, and a proportionality coefficient of $\alpha_2 = 5 \times 10^{-4}$ is selected for the stiffness matrix. The damping matrix **C** is formulated as follows.

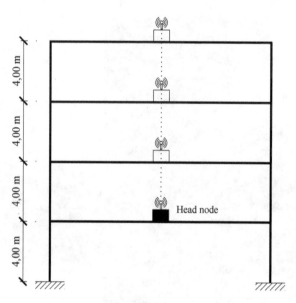

Fig. 10.3 Model of the four-story shear frame structure

Table 10.1 Properties of the shear frame structure

Floor (i)	Story mass (m_i) (kg)	Col. sections ($t_z \times t_y$) (cm × cm)	Column stiffness (k_i) (kN/m)	Number of columns per story	Story stiffness (k_i) (kN/m)
1	20 × 10³	60 × 60	37,939.11	2	75,878.22
2	15 × 10³	40 × 50	14,925.37	2	29,850.75
3	12 × 10³	30 × 50	11,194.03	2	22,388.06
4	10 × 10³	25 × 50	9,328.36	2	18,656.72

$$\mathbf{C} = \alpha_1 \cdot \mathbf{M} + \alpha_2 \cdot \mathbf{K} \qquad (10.4)$$

The acceleration response data of the shear frame structure is derived from time history analysis (test 1). More specifically, a finite element model of the shear frame structure is created, using the SAP2000 finite element software package (Computers and Structures, Inc. 2000). The finite element model is subjected to free vibration after a static load is applied at the top story level and then removed to deflect the shear frame structure from the equilibrium position. Acceleration response data at the mid-span of each story is collected and processed by the sensor nodes of the SHM system.

As mentioned previously, noise may affect the automated on-board calculations of velocities and displacements causing spurious results. Therefore, prior to being processed by the embedded Java classes of the sensor nodes, the acceleration response data is contaminated with artificial noise. The simulated noise is assumed to be "white", i.e. following a Gaussian distribution with zero mean and unity standard deviation. Following observations from preliminary analyses of the effect of noise to the performance of the SHM system, the maximum value of the added noise is set to 1 mg, as meaningful results of the stiffness values could only be calculated for noise levels equal to or lower than 1 mg.

The time step of the time history analysis is set to $\Delta t = 0.01$ s, representing a sampling rate of the sensor nodes of 100 Hz. The sensor nodes in the simulation, as shown in Fig. 10.3, are placed at the mid-span of each story, and the acceleration response data is contaminated with artificial noise before being processed by the embedded software of the sensor nodes. A total of 4,096 acceleration response data is collected from each floor. The software is launched and set to model updating mode and, using the acceleration response data previously contaminated, the calculation of velocities and displacement is performed on each node.

The Newmark-β algorithm used for the integration of the acceleration response data is given in the following equations.

$$\dot{u}_{n+1} = \dot{u}_n + \Delta t (1 - \gamma) \ddot{u}_n + \Delta t \gamma \ddot{u}_{n+1} \tag{10.5}$$

$$u_{n+1} = u_n + \dot{u}_n \Delta t + \Delta t^2 \left(\frac{1 - 2\beta}{2} \right) \ddot{u}_n + \Delta t^2 \beta \ddot{u}_{n+1} \tag{10.6}$$

where n denotes a discrete point of the time history, $\ddot{u}_n$ is the acceleration, $\dot{u}_n$ is the velocity, u_n is the displacement, Δt is the time step, and γ and β are integration coefficients. The values of the coefficients of Eqs. 10.5 and 10.6 are set to $\gamma = 0.5$ and $\beta = 0.25$ following the recommendations of Newmark (1959).

As described earlier, velocity and displacement data are transformed into the frequency domain. The Fourier amplitudes corresponding to spectral peaks of the modes of vibration of the structure are exchanged between neighboring wireless sensor nodes, and one system of dynamic equilibrium equations is formulated on each sensor node. Then, Eq. 10.3 is solved by the on-board application, and the stiffness and damping values are calculated. For the example shown in Fig. 10.3, the actual stiffness, damping, and mass matrices of the FEM model of the entire structure, constructed after assembling the stiffness and damping matrices of each substructure, are:

$$\mathbf{K}_{\mathbf{FEM}} = \begin{bmatrix} (k_1 + k_2) & -k_2 & 0 & 0 \\ -k_2 & (k_2 + k_3) & -k_3 & 0 \\ 0 & -k_3 & (k_3 + k_4) & -k_4 \\ 0 & 0 & -k_4 & k_4 \end{bmatrix} \tag{10.7}$$

$$\mathbf{K}_{\mathbf{FEM}} = \begin{bmatrix} 105,728.97 & -29,850.75 & 0 & 0 \\ -29,850.75 & 52,238.81 & -22,388.06 & 0 \\ 0 & -22,388.06 & 41,044.78 & -18,656.72 \\ 0 & 0 & -18,656.72 & 18,656.72 \end{bmatrix} \tag{10.8}$$

$$
\mathbf{M_{FEM}} = \begin{bmatrix} m_1 & 0 & 0 & 0 \\ 0 & m_2 & 0 & 0 \\ 0 & 0 & m_3 & 0 \\ 0 & 0 & 0 & m_4 \end{bmatrix} = \begin{bmatrix} 20 & 0 & 0 & 0 \\ 0 & 15 & 0 & 0 \\ 0 & 0 & 12 & 0 \\ 0 & 0 & 0 & 10 \end{bmatrix} \tag{10.9}
$$

$$
\mathbf{C_{FEM}} = 0.5 \cdot \mathbf{M_{FEM}} + 5 \cdot 10^{-4} \cdot \mathbf{K_{FEM}} = \begin{bmatrix} 15.29 & -1.49 & 0 & 0 \\ -1.49 & 10.11 & -1.12 & 0 \\ 0 & -1.12 & 8.05 & -0.93 \\ 0 & 0 & -0.93 & 5.93 \end{bmatrix} \tag{10.10}
$$

The stiffness and damping parameters are calculated following the model updating stage introduced above. By solving Eq. 10.3 and assembling the stiffness matrices of all substructures, the calculated stiffness matrix of the structure is:

$$
\mathbf{K_{calc}} = \begin{bmatrix} 105,896.68 & -29,891.61 & 0 & 0 \\ -29,905.16 & 52,298.81 & -22,406.90 & 0 \\ 0 & -22,392.52 & 41,005.35 & -18,635.05 \\ 0 & 0 & -18,824.12 & 18,788.20 \end{bmatrix} \tag{10.11}
$$

Preliminary analyses have shown that the calculation of damping values is prone to inaccuracies due to the artificial noise being added, so calculated damping values may not be considered reliable to serve as a basis for condition assessment. On the other hand, the stiffness values in Eq. 10.11 are very close to the actual stiffness values of the FEM model shown in Eq. 10.8.

Condition assessment of the shear frame structure

For the second stage of the proposed methodology, the condition assessment, damage is simulated on the shear frame structure to validate the ability of the wireless SHM system to identify damage. Damage is simulated in one of the columns of the first story and in one of the columns of the second story of the shear frame structure, as shown in Fig. 10.4. More precisely, damage is simulated by modifying the stiffness, i.e. by adding a reduction factor to the stiffness parameters. Both the moment of inertia and the shear area of one of the first story columns of the finite element model are reduced to 50% of the respective initial value and another time history analysis is performed with the modified stiffness values (Fig. 10.4).

As mentioned above, the performance of damage detection depends on the residuals of the dynamic equilibrium equations, which are applied using newly collected acceleration response data and the initial structural parameters. Moreover, the accuracy of the calculations is affected by noise and numerical approximations; hence, relatively small residuals are still expected even if there is no damage in the structure. Consequently, an additional analysis (test 2) corresponding to the initial state is necessary for comparison purposes. As in test 1, the acceleration response data obtained in test 2 is contaminated with artificial white noise, which is in

Fig. 10.4 Damage introduced into the shear frame structure

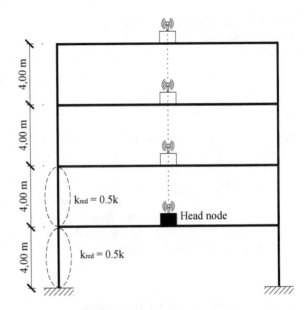

general different than the noise added in test 1. Therefore, small residuals are expected in the results of the additional analysis.

Following the same analysis options as in the model updating stage, a third time history analysis of the finite element model corresponding to the damage scenario is performed (test 3). The acceleration response data at the mid-span of each story is collected and processed by the Java classes embedded into the corresponding sensor node. The sensor nodes use the newly derived data and the initial model parameters to apply the dynamic equilibrium equations. The damage is illustrated in the comparison of the Fourier spectra between the initial state (test 1) and the damaged state (test 3). As shown in Fig. 10.5, the first mode of vibration has been shifted

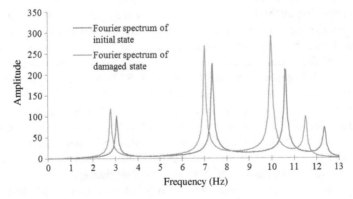

Fig. 10.5 Comparison of acceleration Fourier spectra between test 1 (initial state) and test 3 (damaged state)

Table 10.2 Residuals from dynamic equilibrium equations in test 2 and test 3

n	Equation	Test 2	Test 3
1	$k_{11} \cdot u_1 + k_{12} \cdot u_2 + m_1 \cdot \ddot{u}_1 =$	−23.90	105.45
2	$k_{21} \cdot u_1 + k_{22} \cdot u_2 + k_{23} \cdot u_3 + m_2 \cdot \ddot{u}_2 =$	32.22	1,215.18
3	$k_{32} \cdot u_1 + k_{33} \cdot u_2 + k_{34} \cdot u_3 + m_3 \cdot \ddot{u}_3 =$	−39.17	−765.98
4	$k_{43} \cdot u_1 + k_{44} \cdot u_2 + m_4 \cdot \ddot{u}_4 =$	19.50	−399.30

from $f_1 = 3.05$ Hz to $f_1 = 2.78$ Hz. The residuals of the four dynamic equilibrium equations ($n = 4$) for the first mode ($f_1 = 3.05$ Hz) of the structure (both test 2 and test 3) are summarized in Table 10.2.

It can be concluded from the results of Table 10.2 that the performance of the condition assessment is satisfactory. The deviations from equilibrium in test 2 can be attributed to randomness and approximation errors. However, the deviations from equilibrium in test 3 are considerably larger and, therefore, clearly distinguishable from the deviations of test 2 and thus indicative of damage.

10.4 Summary and Conclusions

Condition assessment of civil infrastructure is an integral part of infrastructure health monitoring. Owing to the significant merits of wireless sensor nodes in terms of cost efficiency and reduced installation time, wireless sensor networks are increasingly employed for infrastructure health monitoring. In this paper, a methodology for decentralized condition assessment of civil infrastructure has been presented. The objective is to enhance the ability of wireless sensor nodes to assess the condition of monitored structures by utilizing embedded computing strategies implemented into wireless sensor nodes. A decentralized numerical model of the monitored structure is embedded into the sensor nodes and the model properties are derived by employing system identification principles in a 2-stage approach. More specifically, using acceleration response data collected by the sensor nodes as well as time integration methods, velocity and displacement data are calculated directly on the sensor nodes. Then, the acceleration, velocity and displacement data is transformed into the frequency domain using the FFT, and the respective Fourier amplitudes of the frequency spectral peaks that correspond to modes of vibration are exchanged between neighboring sensor nodes. Stiffness and damping values are calculated by solving the dynamic equilibrium equations on each node separately for the part of the structure the node is attached to ("model updating" stage). Finally, data derived from an unknown damaged state is used for validation of the damage detection capabilities, applying the dynamic equilibrium equations with the initial model parameters ("condition assessment" stage). Deviations from equilibrium exceeding a predefined threshold could indicate damage.

The validation of the proposed methodology has been performed through simulations on a finite element model of a four-story shear frame structure. In the simulations, one sensor node has been placed on the mid-span of each story. Under free vibration, time history analysis has been performed on the finite element model, in order to collect acceleration response data from the mid-span of each story (test 1). Artificial noise has been added to the data in order to account for the effects of external factors, such as ambient noise. The artificially contaminated acceleration response data from each story has been fed to the software application, written in Java, that has been embedded into the corresponding wireless sensor nodes, and the displacement and velocity data has been automatically calculated. The acceleration, velocity and displacement data has been transformed into the frequency domain. Fourier amplitudes of acceleration, velocity and displacement data have been wirelessly communicated between sensor nodes of neighboring stories (representing substructures), and a system of dynamic equilibrium equations has been formulated on each node. The results of each system of equations, solved by the respective sensor node, has returned the stiffness values of the substructure the node is attached to. From preliminary analyses, it could be observed that, due to numerical approximations and the effects of noise, damping values are particularly prone to large discrepancies from the actual values. Therefore, only stiffness parameters have been deemed suitable for condition assessment.

Damage has been simulated by reducing structural parameters of the columns of the shear frame structure in order to test the system's ability of damage detection. More specifically, the stiffness of one column of the first story and one column of the second story has been reduced. Two tests have been performed, one corresponding to the undamaged state (test 2) and one to the damaged state (test 3). A new set of acceleration response data has been derived at each story and the respective velocities and displacements have been calculated. The newly obtained accelerations, velocities and displacements have been transformed into the frequency domain, and the respective Fourier amplitudes have been used to apply the dynamic equilibrium equations with the initial model parameters. Due to numerical approximations and the interference of ambient noise, it is probable that there are always deviations from equilibrium. To this end, the results of test 3 are compared to the results of test 2, in order to distinguish between deviations attributed to randomness and deviations attributed to damage.

In summary, the methodology has been satisfactory in establishing a reliable numerical model on-board the sensor nodes and in using that model to detect damage in a fully decentralized manner. As shown in this paper, the deviations from equilibrium attributed to damage are clearly distinguished from the deviations that are attributed to randomness. An important point is the establishment of the threshold, below which deviations are not indicative of damage. The threshold can be established by performing preliminary tests on the monitored structure to assess the effects of numerical approximations and of the interference of ambient noise.

Future research may address potential shortcomings of the methodology discussed in this paper. First, further validation tests will be conducted to investigate the stability of the embedded algorithms and the effects of various levels of noise. The extension of the methodology to cover a broad wealth of civil infrastructure systems will also be addressed, while alternative ways to obtain model parameters on a sensor node level will also be investigated.

Acknowledgements Major parts of this work have been developed within the framework of the DeGrie Lab, a Greek-German scientific and educational network funded by the German Academic Exchange Service (DAAD). Also, the authors would like to gratefully acknowledge the support offered by the German Research Foundation (DFG) through the Research Training Group GRK 1462 ("Evaluation of Coupled Numerical and Experimental Partial Models in Structural Engineering"). Any opinions, findings, conclusions or recommendations expressed in this paper are those of the authors and do not necessarily reflect the views of DFG.

References

Brincker R, Andersen P, Zhang L (2000) Modal identification from ambient responses using frequency domain decomposition. In: Proceedings of the 18th international modal analysis conference (IMAC), San Antonio, TX, USA, 07 Feb 2000

Celebi M (2002) Seismic instrumentation of buildings (with emphasis on federal buildings). Technical report No. 0-7460-68170, United States Geological Survey, Menlo Park, CA, USA

Cho S, Yun CB, Lynch JP, Zimmerman A, Spencer Jr B, Nagayama T (2008) Smart wireless sensor technology for structural health monitoring. Steel Struct 8(10.4):267–275

Chowdhury S, Olney P, Deeb M, Zabel V, Smarsly K (2014) Quality assessment of dynamic response measurements using wireless sensor networks. In: Proceedings of the 7th European workshop on structural health monitoring (EWSHM), Nantes, France, 08 July 2014

Cunha Á, Caetano E, Magalhães F, Moutinho C (2005) From input-output to output-only modal identification of civil engineering structures. In: Proceedings of the 1st IOMAC conference, Copenhagen, Denmark, 26 Apr 2005

Dragos K, Smarsly K (2015) A comparative review of wireless sensor nodes for structural health monitoring. In: Proceedings of the 7th international conference on structural health monitoring of intelligent infrastructure. Turin, Italy, 01 July 2015 (Accepted)

Georgieva K, Smarsly K, König M, Law KH (2012) An autonomous landslide monitoring system based on wireless sensor networks. In: Proceedings of the 2012 ASCE international conference on computing in civil engineering, Clearwater Beach, FL, USA, 17 June 2012

Kane M, Zhu D, Hirose M, Dong X, Winter B, Häckel M, Lynch JP, Wang Y, Swartz A (2014) Development of an extensible dual-core wireless sensing node for cyber-physical systems. In: Proceedings of SPIE, sensors and smart structures technologies for civil, mechanical, and aerospace systems, San Diego, CA, USA, 09 March 2014

Law KH, Smarsly K, Wang Y (2014) Sensor data management technologies for infrastructure asset management. In: Wang ML, Lynch JP, Sohn H (eds) Sensor technologies for civil infrastructures. Woodhead Publishing Ltd., Sawston, pp 3–32

Lei Y, Shen WA, Song Y, Wang Y (2010) Intelligent wireless sensors with application to the identification of structural modal parameters and steel cable forces: from the lab to the field. Adv Civil Eng 2010:1–10

Lynch JP, Sundararajan A, Law KH, Sohn H, Farrar CR (2004) Design of a wireless active sensing unit for structural health monitoring. In: Proceedings of SPIE's 11th annual international symposium on smart structures and materials, San Diego, CA, USA, 14 March 2004

Newmark NM (1959) A method of computation for structural dynamics. J Eng Mech-ASCE 85 (10.3):67–94

Peeters B, De Roeck G (1999) Reference-based stochastic subspace identification for output-only modal analysis. Mech Sys Signal Pr 13(10.6):855–878

Rice JA, Mechitov K, Sim SH, Nagayama T, Jang S, Kim R, Spencer BF Jr, Agha G, Fujino Y (2010) Flexible smart sensor framework for autonomous structural health monitoring. Smart Struct Syst 6(5–6):423–438

SAP2000 (2000). Structural analysis program, Computers and Structures, Inc., Berkeley, CA, USA

Smarsly K (2013) Agricultural ecosystem monitoring based on autonomous sensor systems. In: Proceedings of the 2nd international conference on agro-geoinformatics. Center for spatial information science and systems, George Mason University, Fairfax, VA, USA, 12 Aug 2013

Smarsly K (2014) Fault diagnosis of wireless structural health monitoring systems based on online learning neural approximators. In: International scientific conference of the Moscow state university of civil engineering (MGSU), Moscow, Russia, 12 Nov 2014

Smarsly K, Georgieva K, König M (2014) An internet-enabled wireless multi-sensor system for continuous monitoring of landslide processes. Int J Eng Technol 6(10.6):520–529

Smarsly K, Georgieva K, König M, Law KH (2012) Monitoring of slope movements coupling autonomous wireless sensor networks and web services. In: Proceedings of the 1st international conference on performance-based life-cycle structural engineering, Hong Kong, China, 05 Dec 2012

Smarsly K, Law KH (2012) Coupling wireless sensor networks and autonomous software for integrated soil moisture monitoring. In: Proceedings of the international water association (IWA) 10th international conference on hydro informatics, Hamburg, Germany, 14 July 2012

Smarsly K, Law KH (2013a) A migration-based approach towards resource-efficient wireless structural health monitoring. Adv Eng Inform 27(10.4):625–635

Smarsly K, Law KH (2013b) Decentralized fault detection and isolation in wireless structural health monitoring systems using analytical redundancy. Adv Eng Softw 73:1–10

Smarsly K, Law KH, König M (2011) Autonomous structural condition monitoring based on dynamic code migration and cooperative information processing in wireless sensor networks. In: Proceedings of the 8th international workshop on structural health monitoring, Stanford, CA, USA, 13 Sep 2011

Smarsly K, Petryna Y (2014) A decentralized approach towards autonomous fault detection in wireless structural health monitoring systems. In: Proceedings of the 7th European workshop on structural health monitoring (EWSHM), Nantes, France, 08 July 2014

Strutt JW (3rd Baron Rayleigh) (1877) The theory of sound. Mc Millan and Co. Publications, London, UK

SunSPOT (2007) Sun Small Programmable Object Technology datasheet. Oracle Corporation, Redwood Shores, CA, USA

Timoshenko S (1921) On the correction for shear of the differential equations for transverse vibration of prismatic bars. Phil Mag 41:744–746

Wang Y, Lynch JP, Law KH (2007) A wireless structural health monitoring system with multithreaded sensing devices: design and validation. Struct Infrastruct E 3(10.2):103–120

Wang Y, Schwartz A, Lynch JP, Law KH, Lu KC, Loh CH (2006) Wireless feedback structural control with embedded computing. In: Proceedings of the SPIE 11th international symposium on nondestructive evaluation for health monitoring and diagnostics, San Diego, CA, USA, 26 Feb 2006

Zimmerman A, Lynch JP (2007) Parallelized simulated annealing for model updating in ad-hoc wireless sensing networks. In: Proceedings of the international workshop on data intensive sensor networks (DISN'07), Mannheim, Germany, 01 May 2007

Zimmerman A, Shiraishi M, Schwartz A, Lynch JP (2008) Automated modal parameter estimation by parallel processing within wireless monitoring systems. ASCE J Infrastruct Syst 14(1):102–113

Chapter 11
Effects of Local Site Conditions on Inelastic Dynamic Response of R/C Bridges

Ioanna-Kleoniki Fontara, Magdalini Titirla, Frank Wuttke, Asimina Athanatopoulou, George D. Manolis and Petia S. Dineva

Abstract The purpose of this work is to study the effects of site conditions on the inelastic dynamic analysis of a reinforced concrete (R/C) bridge by simultaneously considering an analysis of the surrounding soil profile via the Boundary Element Method (BEM). The first step is to model seismic waves propagating through complex geological profiles and accounting for canyon topography, layering and material gradient effect by the BEM. Site-dependent acceleration time histories are then recovered along the valley in which the bridge is situated. Next, we focus on the dynamic behaviour of a R/C, seismically isolated non-curved bridge, which is modelled and subsequently analysed by the Finite Element Method (FEM). A series of non-linear dynamic time-history analyses are conducted for site dependent ground motions by considering non-uniform support motion of the bridge piers. All numerical simulations reveal the sensitivity of the ground motions and the ensuing response of the bridge to the presence of local soil conditions. It cannot establish a priori that these site effects have either a beneficial or a detrimental influence on the seismic response of the R/C bridge.

Keywords Bridges · Local site conditions · Hybrid methods · Inelastic effects

I.-K. Fontara · F. Wuttke (✉)
Institute of Applied Geo-Sciences, Christian-Albrecht's University, 24118 Kiel, Germany
e-mail: fw@gpi.uni-kiel.de

F. Wuttke
Chair of Marine and Land Geomechanics and Geotechnics, Kiel University, Ludewig-Meyn St. 10, 24118 Kiel, Germany

F. Wuttke
Faculty Civil Engineering, Formerly, Bauhaus-University Weimar, Coudraystrasse 11C, 99423 Weimar, Germany

A. Athanatopoulou · G.D. Manolis
Department of Civil Engineering, Aristotle University, 54124 Thessaloniki, Greece

M. Titirla · P.S. Dineva
Institute of Mechanics, Bulgarian Academy of Sciences, 1113 Sofia, Bulgaria

© Springer International Publishing AG 2017
A.G. Sextos and G.D. Manolis (eds.), *Dynamic Response of Infrastructure to Environmentally Induced Loads*, Lecture Notes in Civil Engineering 2,
DOI 10.1007/978-3-319-56136-3_11

11.1 Introduction

It is well known that geological irregularities of all types produce local distortions in the incoming wave field. Such distortions are generally known as "site effects", and result in a pronounced spatial variation in seismically-induced ground motions. More specifically, spatial variation in seismic ground motions is manifested as measurable differences in amplitude and phase of seismic motions recorded over extended areas. It has an important effect on the response of lifelines such as bridges, pipelines, communication and power transmission grids, tunnels, etc., because these structures extend over long distances parallel to the ground and their supports undergo differential motions during an earthquake.

Simplifying assumptions regarding contemporary bridge design in Eurocode 8 (2003) state that: (i) seismic motion is transmitted to the structure through its supports and is identical at all piers and abutments and (ii) local site conditions are accounted for in terms of site categorization. However, seismic motions are influenced by the wave propagation path and the surface topography at the site of interest, making them a highly variable design parameter.

Several previous studies indicate that local site conditions can exert a crucial influence on the severity of structural damage. Among them, we mention Sextos et al. (2003a, b), who developed a general methodology for deriving appropriate modified time histories that account for spatial variability, site effects and soil structure interaction phenomena. Parametric analyses were conducted and demonstrated that the presence of site effects strongly influences the input seismic motion and the ensuing dynamic response of the bridge. Jeremic et al. (2009) proposed a numerical simulation methodology and conducted numerical investigations of seismic soil-structure interaction for a bridge structure on non-uniform soil. It was then stated that the dynamic characteristics of earthquakes, soil and structure all play a crucial role in determining the seismic behavior of infrastructure-type projects. Zhou et al. (2010) investigated canyon topography effects on the linear response of continuous, rigid frame bridges under oblique incident SV waves. The seismic response of the canyon was analyzed using the FEM, while the response of the bridge was computed by the large mass method. It was shown that the distribution of ground motions is affected by canyon topographic features and the incident angle of the waves. In case of vertical incident SV waves, the peak ground accelerations increase greatly at the upper corners of the canyon and decrease at the bottom corners of the canyon.

In the above mentioned as well as other related studies, however, with exception of Zhou et al. (2010), the influence of local site conditions is evaluated using models based on a uni-dimensional description of the local soil profile as a soil column and similarly for the seismic wave propagation path. It is evident that there is a lack of high-performance computational tools able to simulate two and possibly three dimensional complex geological profiles. The BEM is nowadays recognized as a valuable numerical technique to solve the problem discussed here, due to many advantages in comparison with domain techniques such as the FEM. We briefly mention here the possibility to deal with semi-infinite media in terms of high accuracy and minimal modeling effort.

The main objective of the present work is to investigate the effects of local soil conditions on the dynamic response of R/C bridges. Briefly, the procedure consists of the following steps: (i) Time history records are considered as an input at the seismic bed of complex geological profiles with canyon topography, soil layering and material gradient effect; (ii) next, site dependent ground motions are generated at the surface using the BEM technique; (iii) these are used as input to a three dimensional, seismically isolated model of an R/C, non-curved bridge; (iv) the bridge is then modeled and analyzed using the FEM; (v) different time records are considered as input at each support point of the bridge; (vi) the dynamic response of the bridge due to site dependent ground motions is determined and the results are interpreted to establish changes in terms of what would be observed for a homogenous soil deposit.

11.2   Seismic Signal Recovery Methodology

The BEM is used to model the seismic wave propagation through complex geological profiles so as to recover ground motion records that account for local site conditions. In particular, consider two dimensional wave propagation in viscoelastic, isotropic and inhomogeneous half-plane consisting of N parallel or non-parallel inhomogeneous layers Ω_n ($n = 0, 1, 2, \dots N$) with a free surface and sub-surface

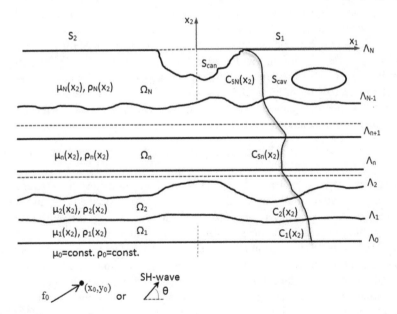

Fig. 11.1 Geometry of the problem treated by BEM: a multi-layered, continuously inhomogeneous geological medium with surface topography and buried inclusions

relief of arbitrary shape. The dynamic disturbance is provided by either an incident SH wave or by waves radiating from an embedded seismic source, see Fig. 11.1. For this problem, a non-conventional BEM is applied which is based on a special class of analytically derived fundamental solution for continuously inhomogeneous media with variable wave velocity profiles (Manolis and Shaw 1996a, b). The employed here BEM was recently developed and validated in Fontara et al. (2015).

More specifically, the material inhomogeneity is expressed by a position-dependent shear modulus and density of arbitrary variation in terms of depth coordinate. We define the inhomogeneity parameter $c_n = C^{bottom\ (\Lambda n-1)}/C^{top\ (\Lambda n)}$ as the ratio of the wave velocity at the bottom to that at the top of any given layer. This model is also able to account for wave dispersion phenomena due to viscoelastic material behaviour and to position-dependent material properties.

Next, for the formulation of the boundary integral equation we use the well-known boundary integral representation formula and insert as kernels the fundamental solutions for geological media with a velocity gradient (Manolis and Shaw 1996a, b).

$$cu_3^{(i)}(\mathbf{x},\omega) = \int_\Gamma U_3^{*(i)}(\mathbf{x},\mathbf{y},\omega)t_3^{(i)}(\mathbf{y},\omega)d\Gamma - \int_\Gamma P_3^{*(i)}(\mathbf{x},\mathbf{y},\omega)u_3^{(i)}(\mathbf{y},\omega)d\Gamma$$

$$\mathbf{x} \in \Gamma = \Omega_i \cup S_{can} \cup S_{cav}$$

(11.1)

In the above, $\mathbf{x}, \mathbf{y}$ are source and field points, respectively, c is the jump term, U_3^* is the fundamental solution for geological media with variable velocity profile, and $P_3^*(\mathbf{x}, \mathbf{y}, \omega) = \mu(x_2)U_3^*(\mathbf{x}, \mathbf{y}, \omega) n_i(\mathbf{x})$ is the corresponding traction fundamental solution, where $i = 1, 2... N$ is the number of layers. The above equation is written in terms of total wave field and expresses the case of incident SH waves. We note that by using this closed form fundamental solution in the BEM technique, only the layer interfaces, as well as the free and sub-surface relief need be discretized.

After discretization of all boundaries with constant (i.e., single node) boundary elements, the matrix equation system is formed below and displacements along the free surface can be computed:

$$[G]\{t\} - [H]\{u\} = \{0\}$$

(11.2)

The above system matrices G and H result from numerical integration using Gaussian quadrature of all surface integrals containing the products of fundamental solutions times interpolation functions used for representing the field variables. They are fully populated matrices of size $M \times M$, where M is the total number of nodes used in the discretization of all surfaces and interfaces, while vectors u and t now contain the nodal values of displacements and tractions at all boundaries.

Finally, the generation of transient signals from the hitherto derived time-harmonic displacements is achieved through inverse Fourier transformation. Note here that both negative and positive values in the frequency, as well as in the time domain, are considered and both real and imaginary values for the response parameter are employed. The aforementioned BEM numerical implementation and production of the final seismic signal is programmed using the MATLAB (2008) software package.

11.3 Geological Profiles

The methodology described in the previous section is now applied to four different hypothetical geological profiles on which the R/C bridge in question is considered to be located, see Fig. 11.2, in order to examine the influence of the following key parameters: (i) canyon topography; (ii) layering; (iii) material gradient. In particular, the site is represented by the following configurations: (a) a homogeneous layer with flat free surface producing a uniform excitation at all support points of the bridge; (b) a homogeneous layer with a valley following the exact canyon geometry in which the bridge is located; (c) a double homogeneous layer deposit as a damped soil column with a valley at the surface; (d) a two-layer damped soil column with the bridge valley at the surface, in which the top layer is continuously inhomogeneous with parameter $c = 1.2$ expressing an arbitrary variation in the wave speed depth profile. The bottom layer is homogeneous and the interface between the first and the second layers is irregular. All geological profiles are overlying elastic bedrock. The soil material properties of these subsoil geological configurations are shown in Table 11.1.

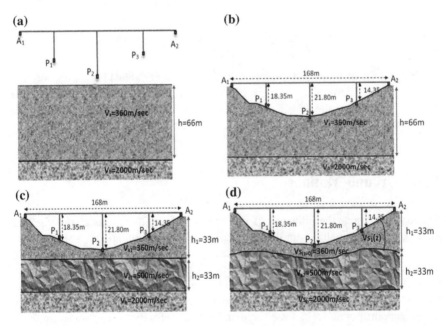

Fig. 11.2 Four geological profiles, Types a–d, on which the R/C bridge is assumed to be located

Table 11.1 Material properties of the basic geological structure

	Vs (m/s)	μ (Pa)	ρ (N/m²)
Layer 1	360	233.28×10^6	1800
Layer 2	500	450×10^6	1800
Bedrock	2000	800×10^7	2000

Table 11.2 Ground motion records from the PEER (2003) strong motion database as recorded on a Class A site

No	Date	Earthquake name	Magnitude (M)	Station name	Closest distance (km)	Component (deg)	PGA (g)
1	22.03.1922	San Francisco	5.3	Golden Gate Park	–	100	0.112
2	17.01.1994	Northridge 1	6.7	Mt Wilson CIT	26.8	000	0.234
3	17.01.1994	Northridge 2	6.7	Littlerock Brainard Can	46.9	090	0.072
4	17.01.1994	Northridge 3	6.7	Lake Hughes #9	28.9	090	0.217
5	18.10.1989	Loma Prieta	6.9	Monterey City hall	44.8	000	0.073
6	10.01.1987	Whittier Narrows	6	Mt Wilson CIT	21.2	000	0.158
7	12.09.1900	Lytle Creek	5.4	Cedar Springs, Allen Ranch	20.6 (Hypocentral)	095	0.071

Next, a suite of seven earthquake excitations given in Table 11.2 are considered, recorded at the outcropping rock on a Class A site according to FEMA classification and are drawn from the PEER (2003) strong motion database. These records are considered as an input at the seismic bed level for all geological profiles.

11.4 Ground Motions

We next investigate the influence of site effects on ground motions recorded along the free surface and start with the first geological profile comprising a single layer with a horizontal free surface that produces a uniform excitation pattern ridge as a reference case. Next, Fig. 11.3 plots the acceleration response spectra recorded at the surface of the Type B geological profile at the support points of the bridge at the canyon and for two different seismic motions. We observe that spectral values at the bridge support points are not the same and furthermore, they differ significantly at certain period values from those produced for the reference case of uniform

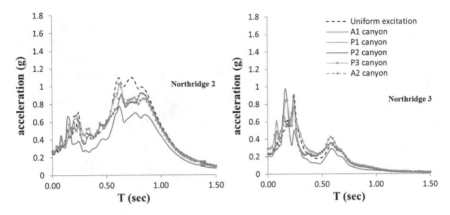

Fig. 11.3 Acceleration response spectra recorded at the free surface of Type B geological profile

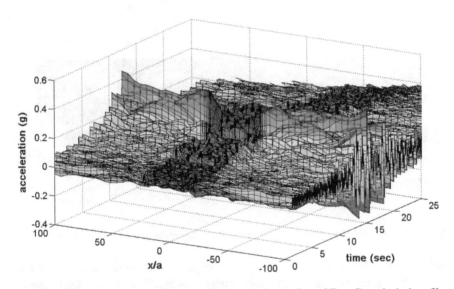

Fig. 11.4 3D acceleration time history recorded at the free surface of Type B geological profile

excitation. Spectral accelerations are more pronounced for low values of period at the bottom of the canyon, while high period values lead to significant spectral acceleration at the edge of the canyon. Three dimensional time history recordings along the canyon are shown in Fig. 11.4, where it is obvious that the seismic signal depends strongly on the canyon topography.

We next examine the influence of canyon topography and of the soil layering on ground motions by comparing acceleration response spectra generated from the uniform excitation geological profile with those generated at the surface of the Type C geological profile that accounts for canyon topography and layering effect,

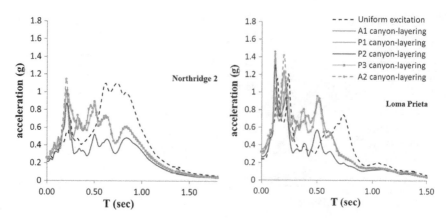

Fig. 11.5 Acceleration response spectra recorded at the free surface of Type C geological profile

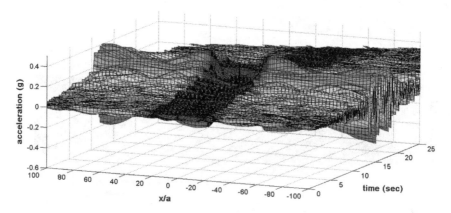

Fig. 11.6 3D acceleration time history recorded at the free surface of Type C geological profile

see Fig. 11.5. We can see that the ground motions are strongly affected by the combined soil layering and canyon topography structure. The shape of the response spectra is now modified, while an expected shifting to the right (higher periods) due to the layering effect is clearly depicted. This is also evident from the 3D time history recorded along the surface of the Type C geological profile shown in Fig. 11.6. There, the acceleration peaks become smoother due to the increased stiffness of the bottom layer.

The combined influence of canyon topography, layering and material gradient effect on the ground motions is now examined. As previously mentioned, in this case the top layer has a continuous variation of the wave speed with depth, avoiding this way the great wave speed contrast between the first and the second layers of the previously examined case, as shown in Fig. 11.7. In addition, we also introduce here a spatial irregularity in the interface between the two soil layers. More specifically, in Fig. 11.8 we compare the acceleration response spectra generated

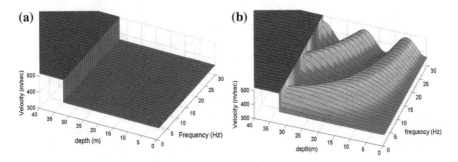

Fig. 11.7 Velocity distribution of the subsoil structure: **a** Type C and **b** Type D geological profile

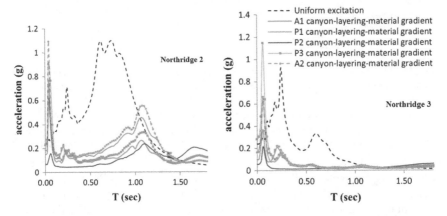

Fig. 11.8 Acceleration response spectra recorded at the free surface of Type D geological profile

for the reference case of uniform excitation with those spectra generated at the surface of the Type D geological profile. We clearly observe now how site effects significantly influence the seismic ground motions. The presence of material gradient increases the material stiffness gradually and the soil becomes stiffer. As a result, the spectral acceleration values are de-amplified across the entire range of periods examined herein.

11.5 R/C Bridge Modeling

We now focus on the nonlinear response of an existing R/C bridge. In particular, we consider the redesign scheme of the Greek Railway Organization (OSE) bridge located in Polycastro, Northern Greece (see Mitoulis et al. 2014). It is a seismically isolated, straight bridge with earthquake resistant abutments and a total length of 168 m supported on rectangular hollow piers of unequal height that varies from 14.35 to 21.8 m, as shown in Fig. 11.9.

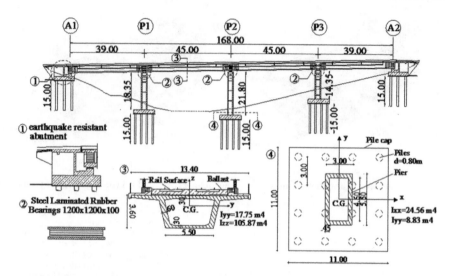

Fig. 11.9 Section details along the bridge span: *1* Longitudinal section of the abutment; *2* steel laminated rubber bearings; *3* deck cross-section; *4* plan view of the pier and its foundation

In terms of some additional details, the two end spans are 39 m long, while the two intermediate spans have span lengths of 45 m each. The concrete deck is a hollow box girder with a constant cross section along the length. For the design of the expansion joints, 40% of the seismic movements of the deck are considered according to Eurocode 8, Part 2 (2003), as well as serviceability-induced constrained movements of creep, shrinkage, pressing and 50% of the thermal movements of the deck. The cracked flexural stiffness of the piers is estimated as equal to 65% of the original cross-section. The fundamental period of the bridge along the *X*-axis is $T_x = 1.43$ s and along the *Y-axis* is $T_y = 0.71$ s.

The bridge is modeled and subsequently analyzed using the FEM commercial program SAP (2007). For modeling the bearings, a number of *N-link* elements are used in order to reproduce the translational and rotational stiffness of the bearings. Piers and deck are modeled by frame finite elements. The flexibility of the foundation of the piers and of the abutments was modeled by assigning six spring elements at the contact points, namely three translational and three rotational ones. These soil spring values were obtained by the geotechnical in situ tests conducted during the final design of the actual bridge. Gap elements are used to model the 25 mm opening at the expansion joints, which separate the backwall from the deck. Note here that the nonlinear response of the bridge is localized and considered only by the non-linearities of the gap elements and of the isolators.

Next, a series of Nonlinear Time History Analyses (NRHA) are conducted under the following conditions: (i) a suite of ground motions applied uniformly to all support points of the bridge and (ii) the same suite of site dependent ground motions, which are now different for each support point of the bridge. These motions also account for (a) canyon topography effect; (b) canyon plus soil layering effect and (c) canyon topography, layering with irregular interfaces and a material gradient effect.

11.6 Dynamic Response of the R/C Bridge

The influence of site effects on structural response of the R/C bridge is demonstrated in Figs. 11.10, 11.11, 11.12 and 11.13 where the input is ground motions at rock outcrop that have been filtered by the complex soil deposits of Fig. 11.2 so as

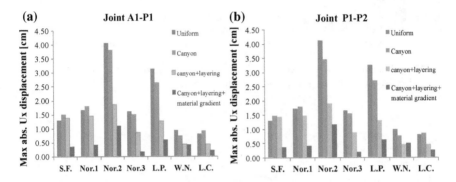

Fig. 11.10 Maximum absolute deck displacements at joints **a** A1–P1; **b** P1-P2, due to ground motions recorded at the surface of the Types A–D geological profiles

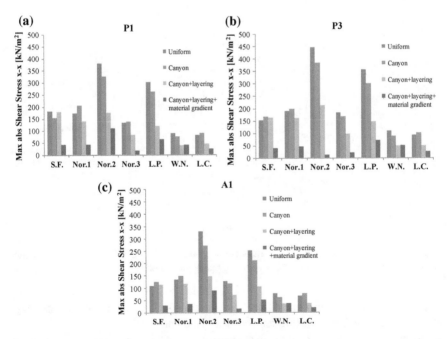

Fig. 11.11 Maximum absolute bearing shear stresses at bearing **a** P1; **b** P3; **c** A1, due to ground motions recorded at the Types A–D geological profiles

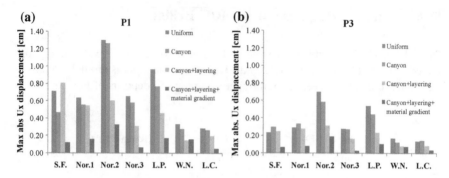

Fig. 11.12 Maximum absolute pier displacements at joint **a** P1; **b** P3, due to ground motions recorded at the surface of the Types A–D geological profiles

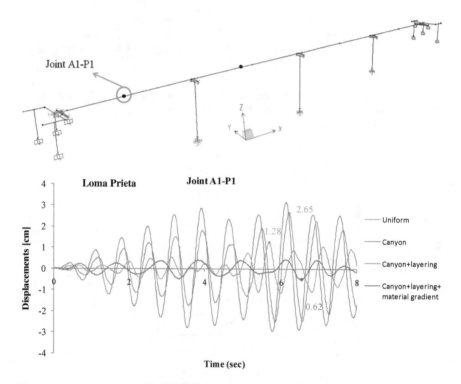

Fig. 11.13 Displacement time history recorded at point A1–P1 on the deck due to ground motions recorded at the surface of the Types A–D geological profiles

to account for (i) uniform excitation, (ii) canyon effect, (iii) canyon and layering effect and (iv) canyon, layering and material gradient effect.

More specifically, maximum displacements of the bridge deck are shown in Fig. 11.10 for the middle point of the first joint (A1–P1) and for the second joint

(P1–P2) along the bridge span for the four types of geological profiles. In most cases, the modified ground motions due to local site effects play an important role in modifying the kinematic response of the bridge in a way that is considered as beneficial. Moving on to the stress field that develops in the R/C bridge, Fig. 11.11 gives the maximum absolute shear stresses at the bearings located at abutment A1 and at piers P1 and P2. We observe that for some seismic motions, local site conditions have a significant effect on the response of the bearings, while for other ground seismic motions local site conditions produce a minor small differences in the bearing response as compared with the reference Type A soil deposit. For at least three ground motions histories, canyon topography results in motions that subsequently overstress the aforementioned supports. However, the input of identical motions as excitations at the bridge's supports will not always yield what is construed as a conservative response, indeed for the examined here case canyon topography effect can lead to 15% increase on the bearings response. Next, maximum absolute displacements at piers P1 and P3 are shown in Fig. 11.12, where we observe that pier P1 is the one most affected by the influence of local soil conditions. For all the cases examined here, ignoring site effects may introduce amplification effects reaching up to 70% in terms of the displacements.

Displacement time histories of the bridge deck, and in particular of the middle point of the first span, are presented in Fig. 11.13 for Loma Prieta ground motion (listed in Table 11.2) due to uniform excitation case and to the ground motions that account for canyon, soil layering and material inhomogeneity. It is observed that canyon topography effect may either amplify or de-amplify the displacement time history of the deck (the latter holds for the present case). When the canyon effect is combined with soil layering, the effect produces strong de-amplification, due to the increase in stiffness of the soil system, plus a shifting of the peaks. The combination of canyon, layering and material gradient effect significantly modifies the

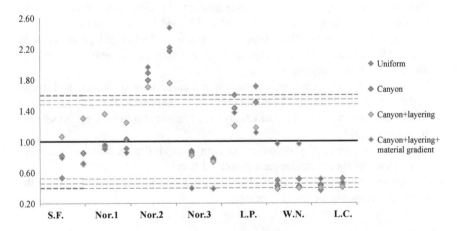

Fig. 11.14 Maximum and minimum normalized deck displacement versus standard deviation for the input ground motions and the four types of geological profiles

displacement time history computed at the deck in terms of resonance frequencies and amplification levels.

In order to generalize the structural behaviour trends, Fig. 11.14 plots the maximum and the minimum normalized displacements of the deck for each ground motion case and for four geological profiles. Comparison is in terms of a mean value plus the standard deviation. As previously mentioned, we again observe that for some ground motions the structural response is significantly affected by the presence of local site conditions, while for other ground motions structural response is only slightly affected.

11.7 Conclusions

In the present study, the influence of local site conditions on the inelastic dynamic response of an existing reinforced concrete bridge located in Northern Greece is investigated using a 2D analysis of the subsoil configuration. A nonconventional BEM technique is applied in order to recover time history records at the surface of complex geological profiles that account for the following combinations: (i) uniform excitation; (ii) canyon effect; (iii) canyon and layering effect; (iv) canyon layering and material gradient effect. Following that, a series of dynamic analyses of the bridge, accounting for lumped nonlinearity, are conducted under site dependent ground motions provided by the previous development, which consider multiple support excitation. From the numerical simulation results, the following conclusions can be drawn:

- Local site conditions cannot be ignored since they significantly influence the inelastic dynamic response of bridges.
- Site dependent ground motions are generated by a new-developed BEM that can represent wave propagation in complex geological media with variable velocity profile, nonparallel layers, surface relief and buried cavities and tunnels.
- The ground motions and the subsequent response of the R/C bridge are strongly affected by the canyon topography, layering and material gradient effect and this effect is frequency-dependent.
- It is not true from the cases examined herein that ignoring site effects and spatial variability of input motions leads to beneficial results for the R/C bridge. Also, it cannot establish a priory that site effects have beneficial influence on the seismic response of bridges.
- Ignoring site effects may introduce an error around 70% in terms of the kinematic field for the particular case examined herein.
- The presence of canyon topography may introduce an increase of 15% in the kinematic response of the R/C bridge.

References

Eurocode 8 (2003) Design of structures for earthquake resistance, part 1: general rules, seismic actions and rules for buildings, part 2: bridges. European Committee for Standardization, Brussels

Fontara I-K, Dineva P, Manolis G, Parvanova S, Wuttke F (2015) Seismic wave fields in continuously inhomogeneous media with variable wave velocity profiles. Arch Appl Mech (under review)

Jeremic B, Jie G, Preisig M, Tafazzoli N (2009) Time domain simulation of soil-foundation-structure interaction in non-uniform soils. Earthq Eng Struct Dyn 38(5):699–718

Manolis GD, Shaw R (1996a) Harmonic wave propagation through viscoelastic heterogeneous media exhibiting mild stochasticity-I. Fundamental solutions. Soil Dyn Earthq Eng 15:119–127

Manolis GD, Shaw RP (1996b) Harmonic wave propagation through viscoelastic heterogeneous media exhibiting mild stochasticity-II. Applications. Soil Dyn Earthq Eng 15:129–139

MATLAB (2008) The language of technical computing, Version 7.7. The Math-Works, Inc., Natick, MA

Mitoulis SA, Titirla MD, Tegos IA (2014) Design of bridges utilizing a novel earthquake resistant abutment with high capacity wing walls. Eng Struct 66:35–44

PEER (2003) Pacific earthquake engineering research center. Strong Motion Database. http://peer.berkeley.edu/smcat/

SAP (2007) Computer and Structures Inc., SAP 2000 Nonlinear Version 11.03, User's Reference Manual, Berkley, CA

Sextos A, Pitilakis K, Kappos A (2003a) Inelastic dynamic analysis of RC bridges accounting for spatial variability of ground motion, site effects and soil-structure interaction phenomena. Part 1: methodology and analytical tools. Earthq Eng Struct Dyn 32:607–627

Sextos A, Pitilakis K, Kappos A (2003b) Inelastic dynamic analysis of RC bridges accounting for spatial variability of ground motion, site effects and soil-structure interaction phenomena. Part 1: parametric study. Earthq Eng Struct Dyn 32:629–652

Zhou G, Li X, Qi X (2010) Seismic response analysis of continuous rigid frame bridge considering canyon topography effects under incident SV waves. Earthq Sci 23:53–61

Chapter 12
BEM-FEM Coupling in the Time-Domain for Soil-Tunnel Interaction

S. Parvanova, G. Vasilev, P.S. Dineva and Frank Wuttke

Abstract The aim of this work is to develop, verify and implement for the pur-poses of numerical simulation studies an efficient, hybrid time-dependent compu-tational technique based on the boundary element method (BEM) and the finite element method (FEM) for evaluation of the seismic response of a complex soil-structure systems. This way, it is possible to take into consideration the entire seismic wave path from the transient seismic source, through stratified with non-parallel layers half-plane, up to the level of a lined tunnel with complex construction. The hybrid numerical scheme is realized via the direct BEM based on the full space elastodynamic fundamental solutions in either Laplace or Fourier domains; Lubich Operational Quadrature in order to obtain time-dependent stiffness matrix and load vector of the seismically active far-field zone; conventional FEM for describing the near-field soil profile together with a lined tunnel; and finally insertion of the BEM model as a macro-finite element in the FEM commercial

S. Parvanova (✉) · G. Vasilev
Department of Structural Engineering, University of Architecture, Civil Engineering and Geodesy, 1046 Sofia, Bulgaria
e-mail: slp_fce@uacg.bg

G. Vasilev
e-mail: gpekov@gmail.com

P.S. Dineva
Institute of Mechanics, Bulgarian Academy of Sciences, 1113 Sofia, Bulgaria
e-mail: petia@imbm.bas.bg

F. Wuttke
Institute of Applied Geo-Science, Christian-Albrechts-University of Kiel, 24118 Kiel, Germany
e-mail: fw@gpi.uni-kiel.de

F. Wuttke
Faculty Civil Engineering, Formerly, Bauhaus-University Weimar, Coudraystrasse 11C, 99423 Weimar, Germany

F. Wuttke
Chair of Marine and Land Geomechanics and Geotechnics, Kiel University, Ludewig-Meyn St. 10, 24118 Kiel, Germany

© Springer International Publishing AG 2017
A.G. Sextos and G.D. Manolis (eds.), *Dynamic Response of Infrastructure to Environmentally Induced Loads*, Lecture Notes in Civil Engineering 2, DOI 10.1007/978-3-319-56136-3_12

software package ANSYS. The accuracy of this approach in the ensuing verification study is discussed. The presented numerical simulations reveal the complex character of the seismic field in an inhomogeneous and heterogeneous geological medium containing additionally an underground structure.

Keywords Soil-structure system · In-plane wave motion · Seismic response · Hybrid time-dependent BEM-FEM technique · ANSYS

12.1 Introduction

The damages caused by an earthquake is a complex result of the interaction of the following important components: (a) the solid earth system, presented by the seismic source, the inhomogeneous and heterogeneous wave path from the source till the site of observation and the local geological profile with its specific geometry, geotechnical and physical conditions; (b) the engineering structures (buildings, bridges, tunnels, dams, pipelines, etc.), their quality and the soil-structure interaction phenomena; (c) the social, economic and political system, which governs the improving of risk assessment and risk management in seismically active geological regions. When analyzing the seismic behavior of structures, kinematic and inertial effects, associated to soil-structure interaction, affect the dynamic characteristics of the interacting system and influence the ground motion around the structure. Building structures, storage tanks, bridges, buried pipes and culverts, retention systems, tunnels and offshore platforms all experience interactive effects. Tunnels are crucial components of the transportation and utility networks in urban areas. The associated impact in case of earthquake induced damage denotes the importance of proper seismic design especially in seismic prone regions. Strong seismic motion of tunnels during earthquakes may result in a partial separation of the underground structure from the soil. Although, tunnels are less prone to damage in comparison with over-ground structures, underestimating the effect of earthquakes on them may lead to major financial and even fatal damages.

The literature is rich with different type of models describing the seismic behavior of underground structures, see Antes and Spyrakos (1997), Spyrakos (2003), Wolf (1984, 1997), where a comprehensive review of the literature on the subject can be found. Among them are simplified spring-dashpot-mass models for structures in homogeneous soil media which could be treated with analytical or semi-analytical techniques (Wolf 1984, 1997; Gucunski 1996), numerical models presenting by FEM (Karabalis and Beskos 1985; Spyrakos and Beskos 1986), finite difference method (Moczo 1989), BEM (Dominguez 1993; Karabalis and Beskos 1986; Tosecky et al. 2005), scaled boundary finite element method (Wolf 2003) and finally hybrid computational schemes (Fah et al. 1990; Karabalis and Beskos 1985; Spyrakos and Beskos 1986; Panza et al. 2009; Wuttke et al. 2011; Manolis et al. 2015; Vasilev et al. 2015) able to model soil-structure systems with more complex geometrical and mechanical properties. In order to profit from the advantages of different approaches and by evading their respective drawbacks, it seems to be quite promising to develop combined formulations. Among all available models the hybrid ones demonstrate greater efficiency and larger potential to deal with the whole path of the seismic wave starting from the source, propagating through the inhomogeneous and heterogeneous geological media and finally loading dynamically the engineering structure that may lead to its damage or even collapse. It is well known in structural analysis that the major strength of the FEM is its versatility in handling various classes of problems, including those involving nonlinearity, anisotropy and nonhomogeneity, while that of the BEM is its ease in data preparation, accuracy in predicting stress concentrations and its ability to handle problems involving the infinite domain. It is then natural for one to attempt to couple these two methods in an effort to create a BEM-FEM scheme that combines all their advantages and reduces or completely eliminates their drawbacks. The following BEM formulations, for solution of transient elastodynamic problems, are available in the literature: (a) approaches based on the use of time-dependent fundamental solution; (b) based on fundamental solutions defined in the Laplace or Fourier transform domain, with the concurrent use of the corresponding inverse integral transform; (c) based on the elastostatic fundamental solution with the inertial terms treated as external body forces; (d) BEM augmented with the convolution quadrature formula of Lubich (1988) for approximating of the Riemann time convolution integrals that arise when Laplace-domain fundamental solutions are used in lieu of time-domain ones. Note that this formulation leads to a time-stepping numerical scheme, see Schanz (1999). The last technique by using Lubich's quadrature is in any case less sensitive to the used mesh-size in comparison to the classical time-domain BEM, since the solution is time independent, which implicitly means a less sensitivity of the method to the spatial mesh-size, see García-Sánchez and Zhang (2007).

The aim of this work is to develop, verify and insert in a simulation study an efficient hybrid time-dependent BEM-FEM technique for evaluation of the seismic response of a complex soil-structure systems, taking into consideration the whole wave path from the transient seismic source, trough stratified with non-parallel layers half-plane, till a lined tunnel with complex structural properties. The proposed computational tool consists of: (1) direct BEM based on the full space elastodynamic fundamental solution in Laplace and Fourier domain; (2) Lubich Operational Quadrature in order to obtain the stiffness matrix in time domain via approximation of the Riemman convolution integral by a special quadrature formula (Lubich 1988); (3) conventional FEM for the near soil region and the underground structure; (4) insertion of the BEM model of the seismically active far-field soil region as a macro element in the FEM commercial program ANSYS; (5) time-dependent solutions are obtained by linear multistep methods using the corresponding solvers.

12.2   Problem Definition

2D elastodynamic problem for a layered soil-tunnel system in an elastic isotropic
half-plane containing an embedded transient seismic source with a prescribed magnitude f_{oi}, (i = 1, 2) is solved. Plane strain state and respectively in-plane wave
motion in the plane $x_3 = 0$ of a Cartesian coordinate system $Ox_1x_2x_3$ is assumed, see
Fig. 12.1. The only non-zero field quantities are displacement components u_1, u_2 and
stresses σ_{11}, σ_{12}, σ_{22} all dependent on the coordinates (x_1, x_2) and on the time t. The
half-plane is divided into two domains: (a) boundary element domain (BED) de-
scribing semi-infinite homogeneous far-field geological region, containing flat free
surface boundary Γ_{FS} and an embedded line seismic source at a point $\mathbf{x}_0$; (b) finite
element domain (FED) comprising of N non-parallel layers, free surface boundary
$\Gamma_{FS} \cup \Gamma_P$ where Γ_P is the boundary of the free surface relief of arbitrary geometry, and
finally a lined infinite cylindrical tunnel with Lame constants, density and material
damping λ_t, μ_t, ρ_t, ξ_t, correspondingly. The internal and external boundaries of the
tunnel are denoted by Γ_{1t} and Γ_{2t} respectively. The BED material properties are λ_0, μ_0,
ρ_0, ξ_0, while those for the i-th layer with interface boundary Γ_i in the FED are denoted
by λ_i, μ_i, ρ_i, ξ_i, i = 1, 2, 3, ... N. The concentrated at point $\mathbf{x}_0$ seismic body force is
$F_i(\mathbf{x}, t) = f_{oi} g(t)\delta(\mathbf{x}, \mathbf{x}_0)$, where its time function is $g(t)$ and δ is the Dirac delta
function. The dynamic equilibrium equation is

$$\sigma_{ij,j}(x_1, x_2, t) = \rho \frac{\partial^2 u_i(x_1, x_2, t)}{\partial t^2} + F_i(x_1, x_2, t), \qquad (12.1)$$

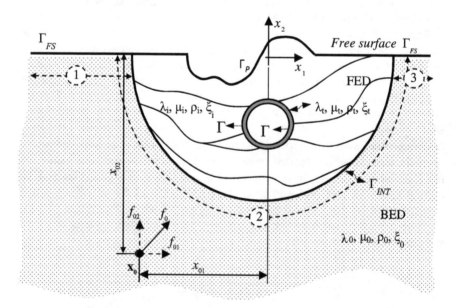

Fig. 12.1 Problem statement

where: $\sigma_{ij} = C_{ijkl}(x_2) u_{k,l}$, $C_{ijkl}(x_2) = \lambda(x_2)\delta_{ij}\delta_{kl} + \mu(x_2)(\delta_{ik}\delta_{jl} + \delta_{il}\delta_{jk})$, δ_{ij} is Kronecker's delta symbol and comma subscripts denote partial differentiation with respect to the spatial coordinates, while the summation convention over repeated indices is implied.

The following boundary conditions are satisfied: (a) inside FED we have zero tractions along boundaries $\Gamma_{FS} \cup \Gamma_P \cup \Gamma_{1t}$ and displacement compatibility and traction equilibrium conditions hold along boundaries $\Gamma_i \cup \Gamma_{2t}$ for $i = 1, 2, \ldots N$; (b) outside FED the tractions along the flat part Γ_{FS} of the free surface belonging to the BED are zero, compatibility and equilibrium conditions hold along interface boundary Γ_{INT} between FED and BED, and finally Sommerfeld radiation condition is satisfied at infinity.

12.3 BEM-FEM Hybrid Technique

The boundary-value problem (BVP), defined above, is solved by hybrid BEM-FEM. The hybrid numerical scheme consists of the following steps:

Step 1 Application of BEM formulation using elastodynamic fundamental solutions in complex domain (Fourier or Laplace) for modeling of infinite BED. This formulation is based on the transformed-domain approach where time dependence is removed by taking a Fourier, Laplace or other integral transform with respect to the time variable. In this case the original hyperbolic partial differential equation of motion (12.1) is reduced to elliptic one that is easier for mathematical modeling. The boundary-value problem in the BED (far-field semi-infinite region with the seismic source) is reformulated via boundary integral equation (12.2) along $\Gamma_{BED} = \Gamma_{FS} \cup \Gamma_{INT}$ with respect to the complex variable z (circular frequency ω in Fourier domain or field variable s in Laplace domain):

$$c_{ij}u_j^{(BED)}(\mathbf{x}, z) = \int_{\Gamma_{BED}} U_{ij}^{*(BED)}(\mathbf{x}, \boldsymbol{\xi}, z)t_j^{(BED)}(\boldsymbol{\xi}, z)dS(\boldsymbol{\xi})$$

$$- \int_{\Gamma_{BED}} P_{ij}^{*(BED)}(\mathbf{x}, \boldsymbol{\xi}, z)u_j^{(BED)}(\boldsymbol{\xi}, z)dS(\boldsymbol{\xi}) \quad (12.2)$$

$$+ f_{0i}\, g(z)U_{ij}^{*(BED)}(\mathbf{x}, \mathbf{x_0}, z).$$

Here: z is frequency ω or Laplace variable s, which are connected by the well-known relation $s = i\omega \rightarrow \omega = -is$, $g(z)$ is Fourier or Laplace transform of the time function, c_{ij} is the jump term depending on the local geometry at the collocation point $\mathbf{x}$, $\mathbf{x}$ and $\boldsymbol{\xi}$ are the position vectors of the source-receiver couple, $t_i = \sigma_{ij}n_j$ are tractions, n_j are the components of the outward normal vector, $U_{ij}^{*(BED)}$ is displacement fundamental solution of the governing equation in complex domain, $P_{ij}^{*(BED)} = C_{iqsl}^{(BED)} U_{sj,l}^{*(BED)} n_q$ is its corresponding traction.

Step 2 The BEM model is converted into macro-finite element by:

(2.1) Condensation of the degrees of freedom of the boundary element model since both FED and BED have only one common contact zone Γ_{INT}. For the aim of degrees of freedom (DOFs) condensation the BED is divided into 3 parts: (a) left flat free surface, part of the Γ_{FS} belonging to the BED, numbered 1; (b) common interface Γ_{INT} denoted 2; (c) right flat free surface, part of the Γ_{FS}, which falls into BED, numbered 3, see Fig. 12.1. As a result relation (12.3) is obtained between traction and displacement vectors, $\mathbf{t}_{INT}$ and $\mathbf{u}_{INT}$ respectively, along Γ_{INT}, taking into consideration the zero traction boundary conditions at the free surface belonging to the BED:

$$\mathbf{t}_{INT} = \mathbf{B}\mathbf{u}_{INT} + \mathbf{P}, \tag{12.3}$$

here the matrix $\mathbf{B}$ of block size 1×1 is equal to $(\mathbf{G}_{22} - [\mathbf{H}_{21} \quad \mathbf{H}_{23}]\mathbf{A}_t)^{-1}([\mathbf{H}_{21} \quad \mathbf{H}_{23}]\mathbf{A}_u + \mathbf{H}_{22})$, where $\mathbf{G}_{ij}$ and $\mathbf{H}_{ij}$ are the well-known influence submatrices along the boundary Γ_{BED} obtained after discretization of Eq. (12.2). Subscripts denote boundary (1 for the left flat surface, 2 for the interface contour, 3 for the right flat free surface). Left subscript is associated with source or collocation point contour, the right one is for receiver or field point boundary. Matrices $\mathbf{A}_t$ and $\mathbf{A}_u$ of block size 2×1 are equal to $\mathbf{A}_t = \begin{bmatrix} \mathbf{H}_{11} & \mathbf{H}_{13} \\ \mathbf{H}_{31} & \mathbf{H}_{33} \end{bmatrix}^{-1} \begin{bmatrix} \mathbf{G}_{12} \\ \mathbf{G}_{32} \end{bmatrix}$

and $\mathbf{A}_u = -\begin{bmatrix} \mathbf{H}_{11} & \mathbf{H}_{13} \\ \mathbf{H}_{31} & \mathbf{H}_{33} \end{bmatrix}^{-1} \begin{bmatrix} \mathbf{H}_{12} \\ \mathbf{H}_{32} \end{bmatrix}$. The vector $\mathbf{P}$ from Eq. (12.3) is

$$\mathbf{P} = (\mathbf{G}_{22} - [\mathbf{H}_{21} \quad \mathbf{H}_{23}]\mathbf{A}_t)^{-1}\left([\mathbf{H}_{21} \quad \mathbf{H}_{23}]\mathbf{\Psi} + \mathbf{\Phi}_2 - [\mathbf{G}_{21} \quad \mathbf{G}_{23}]\begin{Bmatrix} \mathbf{t}_1 \\ \mathbf{t}_3 \end{Bmatrix} \right),$$

where $\mathbf{\Psi} = \begin{bmatrix} \mathbf{H}_{11} & \mathbf{H}_{13} \\ \mathbf{H}_{31} & \mathbf{H}_{33} \end{bmatrix}^{-1}\left(-\begin{Bmatrix} \mathbf{\Phi}_1 \\ \mathbf{\Phi}_3 \end{Bmatrix} + \begin{bmatrix} \mathbf{G}_{11} & \mathbf{G}_{13} \\ \mathbf{G}_{31} & \mathbf{G}_{33} \end{bmatrix}\begin{Bmatrix} \mathbf{t}_1 \\ \mathbf{t}_3 \end{Bmatrix} \right)$ is a vector of block size 2×1 and the load vector $\mathbf{\Phi}_2$ takes into consideration the last term in Eq. (12.2). Details about this condensation could be found in Vasilev et al. (2015).

(2.2) Derivation of the relationship between the vector $\mathbf{F}_{INT}$ of nodal forces of the FE model and the vector $\mathbf{t}_{INT}$ of the discrete tractions of the BE model along the discretized contact zone Γ_{INT}. Finally Eq. (12.4) is obtained representing the BED, described by the BIE (12.2), as a single macro-finite element:

$$\mathbf{F}_{INT} = \mathbf{K}^{(BED)}\mathbf{u}_{INT} + \mathbf{R}, \tag{12.4}$$

where $\mathbf{K}^{BED}$ is the stiffness matrix in Laplace domain and $\mathbf{R}$ is the generalized load vector of nodal forces dependent on the traction load vector $\mathbf{P}$ (see Vasilev et al. 2015).

Step 3   Formation of the stiffness matrix $\mathbf{K}^{BED}$ in time domain, after BEM solution in Laplace domain, and subsequent application of Lubich Operational Quadrature (Lubich 1988). Derivation of the nodal force load vector $\mathbf{R}$, in time domain, by inverse discrete Fourier transform of the BEM solution in frequency domain.

Step 4   Implementation of the macro-finite element in ANSYS software via the ANSYS user programmable features (UPFs) and coupling of both FE and BE time-dependent models by satisfaction of the nodal compatibility and equilibrium conditions required for all available DOFs on the BED-FED contact surface. As a final result the following system of linear algebraic equations is obtained:

$$\sum_{k=1}^{j} \mathbf{K}_k \mathbf{u}_{j-k+1} = \bar{\mathbf{F}}_j, \quad j = 1 \ldots N, \tag{12.5}$$

where N is the total number of time steps. Here the real dynamic stiffness matrices and the nodal force vectors are correspondingly $\mathbf{K}_k = \mathbf{K}_k^{(BED)} + \mathbf{K}_k^{(FED)}$ and $\bar{\mathbf{F}}_j = -\mathbf{R}_j$. The subscript in these notations designates the discrete time value $(t_k = k\Delta t)$ at which the quantities are calculated. In this system the only unknowns are the components of the displacement vector $\mathbf{u}_j$ at time $t_j = j\Delta t$, since all the displacement components in the previous time steps, included in the discrete sum, are previously calculated. This dependence on the previous time steps and not on the future ones is consequence of the causality of system.

The stiffness matrices $\mathbf{K}_k^{(BED)}$ and load vectors $\bar{\mathbf{F}}_j$ are generated with the authors' code developed in MATLAB, while $\mathbf{K}_k^{(FED)}$ is generated by the ANSYS software.

Step 5   Solution of the algebraic system $\mathbf{Ku} = \mathbf{F}$ in time domain by the corresponding solvers in ANSYS. We note that FEM leads to sparse symmetric positive definite matrices, while BEM based on collocation technique leads to full non-symmetric ones. General FEM-BEM coupling is performed by a large matrix partially sparse with the full non-symmetric blocks. When full method of solution in transient analysis is chosen and ANSYS detects the presence of structure with unsymmetrical and fully populated matrices it automatically chooses a solver for dealing with the arisen unsymmetrical system of equations. The unsymmetrical method, which also uses the full [K] and [M] matrices, is meant for problems where the stiffness and mass matrices are unsymmetrical (for example, acoustic fluid-structure interaction problems).

12.4   Verification Study

The aim of this section is to verify the accuracy of the hybrid method in time
domain proposed herein. We have validated software code for modeling of in-plane
problems in the half-plane with surface relief and sub-surface peculiarities in frequency domain (Parvanova et al. 2013, 2014; Dineva et al. 2014a, b). The accuracy
of the BEM authors' software code is verified in Parvanova et al. (2013, 2014),
where the P- and SV-scattered displacement and stress fields, by a semi-circular
canyon or semi-elliptic hill, and sub-surface cavity, are compared with numerical
and analytical solutions of other authors. In order to use this software code for
verification of the results obtained by hybrid BEM-FEM model, we consider
homogeneous in depth elastic and isotropic half-plane with surface topography in
frequency domain. The relevant BEM solution in time domain is obtained by
inverse discrete fast Fourier transform of the original results.

The numerical example, considered here, is a semi-circular canyon of radius
$a = 7$ m located in an elastic, isotropic and homogeneous half-plane with material
parameters: Poisson's ratio $\nu_0 = 1/3$; shear modulus $\mu_0 = 7 \times 10^5 \text{kN/m}^2$; density
$\rho_0 = 2 \text{t/m}^3$; and hysteretic damping ratio $\xi_0 = 0\%$. For validation purposes the
half-plane is divided into finite (FED) and infinite (BED) domains by fictitious
semi-circular interface of radius $r = 10a$, shown as dashed line in Fig. 12.2a. The
load is presented by an embedded source of magnitude $f_{0i} = (0, -1 \times 10^6 \text{kN})$
located at point $\mathbf{x}_0 = (0, -150)$ below the flat surface, which falls within the infinite
half-plane. The transient concentrated force $F_i(\mathbf{x}, t) = f_{0i}g(t)\delta(\mathbf{x}, \mathbf{x}_0)$ in the source is
defined by time-history function $g(t)$ presented in Fig. 12.3a. This function has
continuous Fourier transform in frequency domain given in Fig. 12.3b.

Two numerical models are considered: (1) reference (benchmark) BEM solution
performed by authors' software based on conventional BEM and frequency
dependent full space fundamental solutions. The BE mesh comprises of 144
quadratic boundary elements, 48 of which located along the canyon contour. The
length of the discretized flat free surface is 280 m or 40a, measured left and right to
the canyon center. The Nyquist frequency is 15 Hz and the frequency range is
divided into 300 equal steps. The BEM results in time domain are obtained by

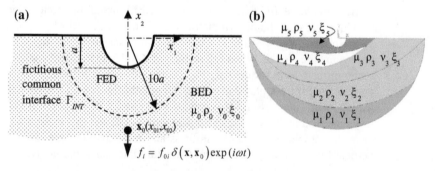

Fig. 12.2 Geometry of the verification example: **a** general view; **b** finite element domain

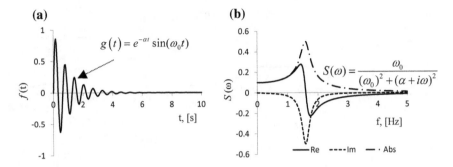

Fig. 12.3 a Time-history function; **b** Fourier transform of g(t)

inverse discrete Fourier transform; (2) The hybrid BEM-FEM model comprises of two domains: infinite homogeneous half plane (BED) and multi-layered FED as shown in Fig. 12.2a and b. The same layer outlines are used in the following simulation study in Sect. 12.5, here, for validation purposes, all layers have material properties identical with those of the half-plane. The FE model comprises of 4618 quadratic FE (ANSYS code PLANE 82) and one macro-finite element (MATRIX 50) with 97 nodes, generated as BED and MATLAB software code. Transient analysis of the FE model is performed, applying Newmark integration scheme, and unknown displacement components of the hybrid model are derived directly in time domain.

Comparisons of the results, obtained by both BEM and hybrid BEM-FEM models, are shown in Fig. 12.4a–d. More specifically Fig. 12.4a depicts vertical displacement components for bottom of the canyon $(\mathbf{x} = (0, -7))$ versus time variable. The plots in Figs. 12.4b–d are free field displacement components for fixed time $t = 0.3667, 0.7$ and 1 s, which corresponds to the first three extreme values of the previous graphics. There, the percentage differences between both solutions are marked for vertical displacements at bottom of the canyon at the relevant time.

The same sets for comparison are plotted in Fig. 12.5 but for shallow seismic source located at point $\mathbf{x}_0 = (50, -21)$, which falls in FED. Here the horizontal displacement at bottom of the canyon is nonzero and is also plotted in Fig. 12.5a. The Nyquist frequency in the BEM model is 15 Hz and the frequency range is divided again into 300 equal steps. Here the time values corresponding to the first three extreme values in the vertical displacements are $t = 0.3, 0.6667$ and 0.9667 s. The percentage differences between both models for vertical displacements at bottom of the canyon are marked at the corresponding site-effect-plots. Such discrepancies cannot be totally avoided, but they are obtained for the most stringent verification test namely the absence of any material damping. We expect that the introduction of material damping in the soil will reduce the percentage differences. Nevertheless, all presented results demonstrate satisfactorily numerical accuracy (less than 5%) between the present hybrid methodology and the results obtained by the pure direct BEM formulation.

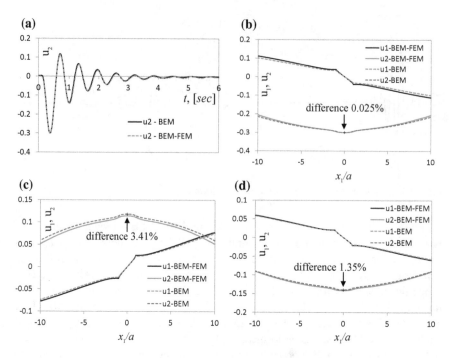

Fig. 12.4 Comparison of the results for line source located at point $\mathbf{x}_0 = (0, -150)$: **a** Vertical displacements at bottom of the canyon $(0, -7)$ versus time; displacement components along the free surface for fixed time: **b** $t = 0.3667$ s; **c** $t = 0.7$ s; **d** $t = 1$ s

12.5 Numerical Simulations

In this section the application of the hybrid technique for modeling of relatively complex soil-structure system is demonstrated. The computational model is an illustrative example of a practical tunnel construction and properties of the soil profile imitate real geotechnical data, although they are not a real case study. Our aim is to show some of the specialties of such models as: (a) site effects, as manifested by the presence of the free-surface relief peculiarities. In the case of the illustrative example we consider a hill; (b) the effect of buried tunnels in the stratified geological media; (c) dynamic stress concentration in the tunnel liners; (d) influence of the near soil profile: homogeneous or layered; (e) successful treatment of infinite half-plane models in time domain by implementing BEM approach in commercial software package.

We consider a multi-story tunnel construction of circular cross section depicted in Fig. 12.6a, which is embedded in a multi-layered FED as shown in Fig. 12.6b. The near soil profile, presented by FED, comprises of 5 geological layers of arbitrary outline, which in case of real practical problem could be reported from the geological profile. The material properties of the adopted soil strata are as follows: $\mu_1 = 620$ MPa, $\nu_1 = 1/3$, $\rho_1 = 1.8$ t/m^3; $\mu_2 = 500$ MPa, $\nu_2 = 0.4$, $\rho_2 = 1.75$ t/m^3;

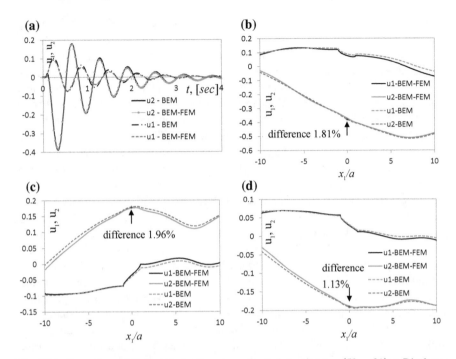

Fig. 12.5 Comparison of the results for line source located at point $\mathbf{x}_0 = (50, -21)$: **a** Displacement components at bottom of the canyon $(0, -7)$ versus time; displacement components along the free surface for fixed time: **b** $t = 0.3$ s; **c** $t = 0.6667$ s; **d** $t = 0.9667$ s

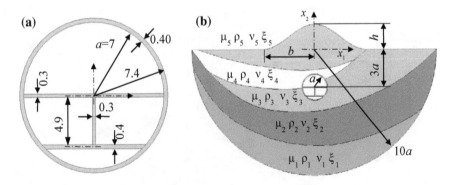

Fig. 12.6 Geometry of the illustrative example: **a** tunnel construction; **b** geological FED strata

$\mu_3 = 200$ MPa, $\nu_3 = 0.45$, $\rho_3 = 1.8$ t/m^3; $\mu_4 = 100$ MPa, $\nu_4 = 0.46$, $\rho_4 = 1.85$ t/m^3; $\mu_5 = 65$ MPa, $\nu_5 = 0.48$, $\rho_5 = 1.9$ t/m^3. The material damping of all layers is assigned as $\xi_1 = \xi_2 = \xi_3 = \xi_4 = \xi_5 = 0\%$. The outline of the FED is delineated by upper traction free surface containing hill with height $h = 14$ m and semi-width $b = 30$ m, and semi-circular lower boundary of radius $10a = 70$ m, which is the

BED-FED common interface. The shape of the hill is described by the following
function $x_2(x_1) = h\cos^2(\pi x_1/2b)$.

The tunnel contains two train lanes on the first level and several travel road lanes
on the upper level. The inner radius of the circular cross section is $a = 7$ m, the
wall thickness is 0.40 m, all dimensions of the tunnel liners and internal plates and
walls are given in Fig. 12.6a. The depth of burial is $h = 3a = 21$ m measured from
the flat free surface. The material properties of all components of the tunnel construction are as follows: modulus of elasticity $E_t = 30000$ MPa, Poisson's ratio
$\nu_t = 0.2$, material density $\rho_t = 2.5\,\text{t/m}^3$, and constant material damping $\xi_t = 0\%$.

The FED itself is embedded in an elastic homogeneous half-plane (BED) with
material parameters $\mu_0 = 700$ MPa, $\nu_0 = 1/3$, $\rho_0 = 2\,\text{t/m}^3$ and material damping
again $\xi_0 = 0\%$, see Fig. 12.7. The load is presented by in-plane waves radiating by
an embedded line transient source of magnitude $f_{0i} = (0, -1 \times 10^6\,\text{kN})$ located at a
point $\mathbf{x}_0 = (0, -150)$ or at a point $\mathbf{x}_0 = (50, -21)$ below the flat surface in BED
and FED respectively. The excitation $F_i(\mathbf{x}, t) = f_{0i}g(t)\delta(\mathbf{x}, \mathbf{x}_0)$ follows the
time-history function $g(t)$ shown in Fig. 12.3. The stiffness matrix and load vector
for the macro–FE, having all features of the infinite half-plane, are generated after
BEM solution in Laplace and frequency domain respectively. The BE model
comprises of 144 quadratic boundary elements, 48 of which located along the
common interface Γ_{INT}. The length of the discretized flat free surface is 2800 m or
$400a$ (40 times radius of the contact contour), for each left and right part to the
contact interface, see Fig. 12.7. The Nyquist frequency is 15 Hz which results in
time step $\Delta t = 1/30$ s. The FEM mesh employed in the numerical disctretization for
the FED comprises 8325 quadratic finite elements (ANSYS code PLANE 82) and

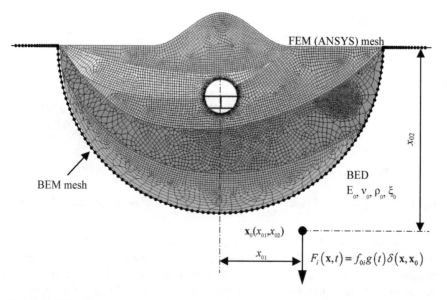

Fig. 12.7 BEM-FEM mesh discretization layouts

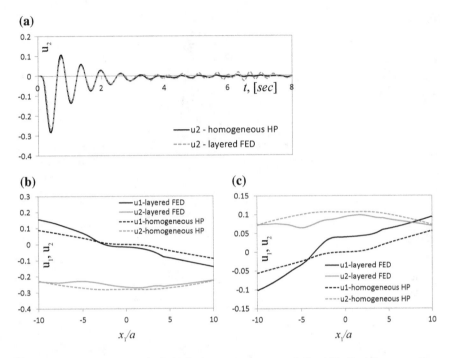

Fig. 12.8 Deep source: **a** Vertical displacements at top of the hill (0, 14) versus time; displacement components along the free surface for fixed time: **b** $t = 0.4$ s; **c** $t = 0.7333$ s

one macro-finite element containing 97 nodes (ANSYS code MATRIX 50). The FE mesh layout and BEM discretization, employed for macro-element formation, are portrayed in Fig. 12.7.

Two numerical models are considered for both the sources in order to point out the influence of the stratified near soil profile: (1) tunnel construction in homogeneous, elastic, and isotropic FED with material properties of the soil deposit identical with those of the half-plane $\mu_0 = \mu_{1-5} = 700$ MPa, $\nu_0 = \nu_{1-5} = 1/3$, $\rho_0 = \rho_{1-5} = 2$ t/m^3 and zero material damping; (2) stratified FED with homogeneous layers and different material properties as reported at the beginning of this section. The first set of results concerns displacement components at the free surface. Figure 12.8a plots vertical displacement components for top of the hill versus time variable in case of deep seismic source at a point $\mathbf{x}_0 = (0, -150)$ and both models: homogeneous half-plane and stratified FED. The surface distribution of both horizontal and vertical displacements, at fixed time $t = 0.4$ s and $t = 0.7333$ s are presented in Figs. 12.8b and c. The chosen time values correspond to the first two extreme values in the vertical displacements of the first plot (Fig. 12.8a). Analogous results are portrayed in Fig. 12.9 but for the case of shallow line source at a point $\mathbf{x}_0 = (50, -21)$. More specifically vertical and horizontal displacement components for top of the hill versus time for both homogeneous and stratified FED are plotted in Fig. 12.9a and b. Site effects as manifested by horizontal and vertical

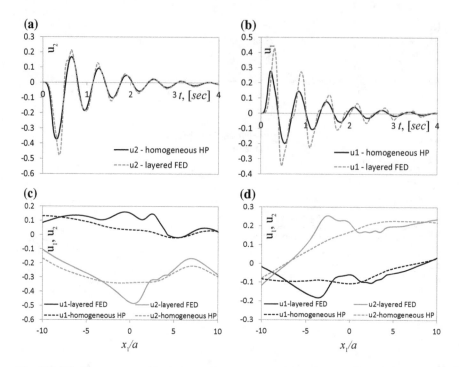

Fig. 12.9 Shallow source: **a** Vertical and **b** horizontal displacements at top of the hill (0, 14) versus time; displacement components along the free surface for fixed time: **c** $t = 0.3667$ s; **d** $t = 0.6333$ s

displacements along the free surface at fixed time $t = 0.3667$ s and $t = 0.6333$ s are presented in Fig. 12.9c and d. Similarly to the deep source the chosen time variables here, correspond to the first two extreme values in the vertical displacements at top of the hill versus time (Fig. 12.9a).

Figures 12.8 and 12.9 illustrate the sensitivity of the synthetic wave field at the free surface to the location of the seismic load and the heterogeneous character of the wave path. As expected in the case of shallow line source horizontal displacements are significant while those due to the deep source are negligible. Contrary, vertical displacement components are prevailing for both the sources (as far as the concentrated force is vertical). Of interest is the fact that the influence of the stratified near soil geological profile is highly pronounced for the shallow source and not so expressive for the deep source. One reason for this picture is the fact that primary P-waves generated from the deep source reach the free surface at both ends of the contact interface propagating mainly in the infinite half-plane and barely passing through the first two layers. That's why the vertical displacements at $x_1/a = \pm 10$ are identical for layered and homogeneous soil profile (Fig. 12.8b and c). Contrary, the transient waves radiated from the shallow source necessarily propagate through the layered soil deposit which results in significant differences in the site effects between both considered models (Fig. 12.9).

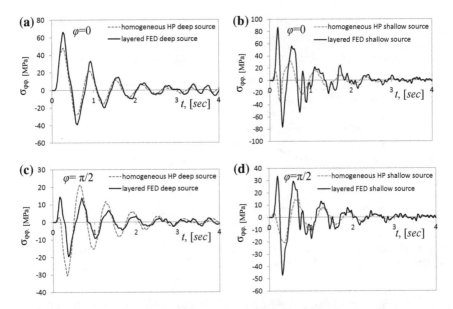

Fig. 12.10 Hoop stresses along the soil-tunnel interface for fixed representative points, from the tunnel side: **a** deep line source and polar angle $\varphi = 0$; **b** shallow line source and $\varphi = 0$; **c** deep source and $\varphi = \pi/2$; **d** shallow source and $\varphi = \pi/2$

The second set of results is for hoop stresses along the soil-tunnel interface, from the tunnel side. Figure 12.10 plots the hoop stresses at two observer points with polar angles $\varphi = 0$ and $\varphi = \pi/2$ versus time for the case of deep line source (a, c) and shallow source (b, d), for both types of soil deposit: homogeneous and layered. Figure 12.11 depicts polar distribution of the hoop stresses at fixed time $t = 0.4$ s and $t = 0.7333$ s and the case of deep line source. Identical stress distributions for the case of shallow source and fixed time $t = 0.3667$ s and $t = 0.6333$ s are portrayed in Fig. 12.12. Here the results obtained by both models are quite different as far as the tunnel is in the core of the stratified FED and all in-plane waves surely propagate through the soil strata in order to reach this interface independently of the source location.

In the case of deep line source polar stress distribution in Fig. 12.11 shows sectors with amplification and deamplification in hoop stresses due to the soil strata. The same picture is observed in Fig. 12.10a and c: for polar angle $\varphi = 0$ layered FED causes amplification, while for $\varphi = \pi/2$ hoop stresses, in case of layered soil profile, are generally reduced compared to those in the case of homogeneous half-plane. In the case of shallow seismic source the hoop stresses are amplified considerably with layering in the soil profile. This statement holds truth for polar distribution in Fig. 12.12 as well as for both representative points in Fig. 12.10b and d.

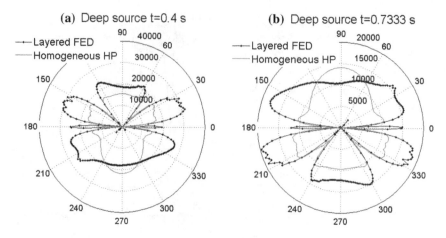

Fig. 12.11 Polar distribution of hoop stresses in kN/m², along the soil-tunnel interface in the case of deep seismic source for a fixed time: **a** $t = 0.4$ s, and **b** $t = 0.7333$ s

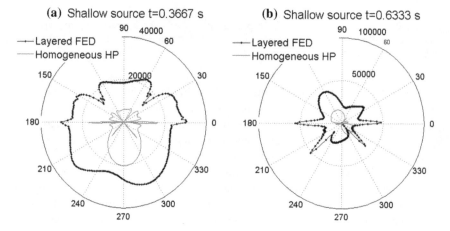

Fig. 12.12 Polar distribution of hoop stresses in kN/m², along the soil-tunnel interface in the case of shallow seismic source for a fixed time: **a** $t = 0.3667$ s, and **b** $t = 0.6333$ s

12.6 Conclusions

The 2D in-plane elastodynamic problem for the seismic response of buried tunnels in a layered half-plane containing a line transient seismic source is solved in time-domain via hybrid BEM-FEM. An efficient coupling strategy is developed which employs the numerical schemes of finite and boundary element methods in a very flexible manner. The proposed hybrid model and the accompanied computational technique has the main advantage that it is possible to describe in one model the time-dependent behavior of the entire system defined by the seismic source with its specific geophysical properties, the inhomogeneous and heterogeneous wave path with complex geometry, the local geotechnical region with free- and sub-surface relief and finally the underground structure with its specific structural properties.

The numerical simulations conducted by the present methodology produced results which show that both scattered wave displacement field on the free surface and the stress concentration field near the lined tunnel with complex structure are sensitive to site conditions such as the existence of surface relief, the presence of layers, source position and most importantly on soil-underground structure interaction effects. In sum, this type of work is useful in the field of earthquake engineering and in the seismic design of structures.

The developed hybrid methodology has the potential to be extended further for treating of both geometrical (large displacements, uplifting phenomena) and physical (inelastic material, unsaturated soils) type of nonlinearities. The idea of spatially subdividing the domain under consideration used here allows successful application of the boundary element methods for unbounded domains, high stress resolution, etc., but most likely more efficiency is gained by using a finite element method in regions where the boundary element method is not required, for example in modeling of the local nonlinearities.

Acknowledgements The first and the second authors wish to acknowledge the support provided by RC&DC of UACEG-Sofia through the Grant No. BN 174/15.

References

Antes H, Spyrakos CC (1997) Soil-structure interaction. In: Beskos DE, Anagnostopoulos SA (eds) Computer analysis and design of earthquake resistant structures: a handbook, Chapter 6. Computational Mechanics Publications, Southampton, pp 271–332

Dineva P, Parvanova S, Vasilev G, Wuttke F (2014a) Seismic soil-tunnels interaction via BEM. I. Mechanical model—(Part I). J Theoret Appl Mech 44(3):31–48

Dineva P, Parvanova S, Vasilev G, Wuttke F (2014b) Seismic soil-tunnels interaction via BEM. II. Numerical results—(Part II). J Theoret Appl Mech 44(4):29–50

Dominguez J (1993) Boundary elements in dynamics. Computational Mechanics Publications, Southampton

Fah D, Suhadolc P, Panza GF (1990) Estimation of strong ground motion in laterally heterogeneous media: Modal summation-finite differences. In: Proceedings of the 9th European conference of earthquake engineering, Moscow, USSR, 11–16 September, vol 4A, pp 100–109

García-Sánchez F, Zhang Ch (2007) A comparative study of three BEM for transient dynamic crack analysis of 2-D anisotropic solids. Comput Mech 40:753–769

Gucunski N (1996) Rocking response of flexible circular foundations on layered media. Soil Dyn Earthq Eng 15(8):485–497

Karabalis DL, Beskos DE (1985) Dynamic response of 3D flexible foundations by time domain BEM and FEM. Soil Dyn Earthq Eng 4(2):91–101

Karabalis DL, Beskos DE (1986) Dynamic response of 3D embedded foundations by the boundary element method. Comput Methods Appl Mech Eng 56:91–119

Lubich C (1988) Convolution quadrature and discretized operational calculus-I. Numer Math 52 (2):129–145

Manolis G, Parvanova S, Makra K, Dineva P (2015) Seismic response of buried metro tunnels by a hybrid FDM-BEM approach. Bull Earthq Eng. doi:10.1007/s10518-014-9698-6

Moczo P (1989) Finite-difference technique for SH waves in 2-D media using irregular grids: application to the seismic response problem. Geophys J Int 99:321–329

Panza G, Paskaleva I, Dineva P, Cr La Mura (2009) Earthquake site effects modelling by hybrid MS-BIEM: the case study of Sofia Bulgaria. Rendiconti di Scienze Fisiche by the Accademia dei Lincei 20:91–116

Parvanova S, Dineva P, Manolis G, Wuttke F (2013) Diffraction of in-plane (P, SV) and anti-plane (SH) waves in a half-space with cylindrical tunnels. In: Sellier A, Aliabadi MH (eds) Proceedings of international conference on boundary element techniques XIV 16–18 July 2013, Paris, France. EC Ltd, UK, pp 335–340. ISBN: 978-0-9576731-0-6

Parvanova S, Dineva P, Manolis G (2014) Elastic wave fields in a half-plane with free Surface relief, tunnels and multiple buried inclusions. Acta Mech 225(7):1843–1865

Schanz M (1999) A boundary element formulation in time domain for viscoelastic solids. Commun Numer Methods Eng 15:799–809

Spyrakos CC, Beskos D (1986) Dynamic response of flexible strip-foundations by boundary and finite elements. Soil Dyn Earthq Eng 5(2):84–96

Spyrakos CC (2003) Soil-structure interaction in practice. In: Oliveto G, Hall WS (eds) Chapter in BEM for soil-structure interaction. Kluwer Academic Publications, London, pp 235–275

Tosecky A, Schmid G, Hackl K (2005) Wave propagation in homogeneous elastic halfspace using the dual reciprocity boundary element method. Ruhr University, Bochum

Vasilev G, Parvanova S, Dineva P, Wuttke F (2015) Soil-structure interaction using BEM-FEM coupling through ANSYS software package. Soil Dyn Earthq Eng. doi:10.1016/j.soildyn.2014.12.007

Wolf JP (1997) Spring-dashpot-mass models for foundation vibration. Earthq Eng Struct Dyn 26:931–949

Wolf JP (1984) Dynamic soil-structure interaction. Prentice-Hall Inc, Englewood Cliffs

Wolf JP (2003) The scaled boundary finite element method. Wiley, England

Wuttke F, Dineva P, Schanz T (2011) Seismic wave propagation in laterally inhomogeneous geological region via a new hybrid approach. J Sound Vib 330:664–684

Chapter 13
Energy Methods for Assessing Dynamic SSI Response in Buildings

Mourad Nasser, George D. Manolis, Anastasios G. Sextos, Frank Wuttke and Carsten Könke

Abstract An energy approach is employed here for assessing model quality for dynamic soil-structure interaction (SSI) analysis. Concurrently, energy measures are introduced and calibrated as general indicators of structural response accuracy. More specifically, SSI models built at various abstraction levels are investigated according to various coupling scenarios between soil and structure. The hypothesis of increasing model uncertainty with decreasing complexity is first investigated. A mathematical framework is provided, followed by a case study involving alternative models for incorporating SSI effects. During the evaluation process, energy measures are used in conjunction with the *adjustment factor approach* to quantify SSI model uncertainty. Two types of uncertainty are considered, namely in the numerical model and in the model input parameters. Investigations on model framework uncertainty show that the 3D finite element (FE) model yields the best quality results, whereas the Wolf

M. Nasser
GERB Schwingungsisolierungen GmbH & Co. KG, 13407 Berlin, Germany

G.D. Manolis (✉) · A.G. Sextos
Department of Civil Engineering, Aristotle University of Thessaloniki,
54124 Thessaloniki, Greece
e-mail: gdm@civil.auth.gr

A.G. Sextos
e-mail: asextos@civil.auth.gr; a.sextos@bristol.ac.uk

F. Wuttke
Faculty of Engineering, Christian-Albrechts University, 24118 Kiel, Germany

C. Könke
Department of Civil Engineering, Bauhaus University, 99423 Weimar, Germany

A.G. Sextos
Department of Civil Engineering, University of Bristol, Bristol BS8 1TR, UK

F. Wuttke
Faculty Civil Engineering, Formerly, Bauhaus-University Weimar,
Coudraystrasse 11C, 99423 Weimar, Germany

F. Wuttke
Chair of Marine and Land Geomechanics and Geotechnics,
Kiel University, Ludewig-Meyn St. 10, 24118 Kiel, Germany

© Springer International Publishing AG 2017   237
A.G. Sextos and G.D. Manolis (eds.), *Dynamic Response of Infrastructure to Environmentally Induced Loads*, Lecture Notes in Civil Engineering 2,
DOI 10.1007/978-3-319-56136-3_13

lumped parameter model produces the lowest model uncertainty in the simple model category. Also, the fixed-base model produces the highest estimated uncertainty and consequently is the worst quality model. The present results confirm the hypothesis that increasing model uncertainty comes with decreasing complexity, but only when the assessment is based on an energy measure as the response indicator.

Keywords Energy methods · Soil-structure Interaction · Earthquake Engineering

13.1 Introduction and Problem Outline

Uncertainties are inherent in all numerical modeling processes and cannot be eliminated. Uncertain model response is generally associated with both input variables and with a priori hypotheses regarding model configuration. More specifically, simplified models provide a rather undemanding framework for structural design, but may be inadequate because idealizations are inconsistent with the actual configuration of complex, asymmetric buildings (Naiem 2001; Nasser et al. 2010). On the other hand, discrete parameter models are more general and incorporate many structural details. However, the computational effort increases considerably when a large number of degrees-of-freedom (DOF) is required to model this level of sophistication, and culminates with the implementation of nonlinear constitutive models for tracing post-elastic behavior. Such large scale models inevitably give rise to uncertainties regarding the selection of values for the input parameters. An energy approach can serve as a powerful tool for the purpose of model assessment, since it is based on a clear physical concept. Of the different parts comprising earthquake-induced energy, the hysteretic energy that structures dissipate is associated with inelastic behavior and consequently with structural damage. Although an increase in the duration of the excitation signal leads to increased input and hysteretic energies, it does not influence their ratio (Rahnama and Manuel 1996).

Pearl (1978) noted that simpler models are preferable and when implemented, they are often considered to be more plausible and can easily be tested. This point of view is also supported by Beck (2006), who pointed out that models of different makeup serve different functions in the field of environmental simulation and that the degree of difficulty in validating models is proportional to their complexity. Oreskes et al. (1994) argued that there is no proof that accurate results are more likely to be obtained from simpler models as compared to more complex models. The idealizations and assumptions made in reference to the various models cannot guarantee results that conform with reality, as construed by experimental testing. Furthermore, Beck (2006) showed that validated models do not automatically produce precise results for different types of applications, as compared to what actually happens in reality. In the literature, we see reference to model framework uncertainty described as either model error or as model uncertainty (EPA 2009; Luis and McLaughlin 1992). This is precisely the type of uncertainty we propose to investigate here. By way of contrast, Beck (2006) uses structural error to express this type of uncertainty, which is associated with the process of creating algorithms that describe and solve structural-type models. Konikow and Bredehoeft (1992) finally refer to model framework uncertainty in terms of conceptual error, which is understood to describe uncertainty in transforming reality into equations that represent and govern a physical system's behavior.

Our aim is to introduce a generic methodology for assessing the quality of dynamic SSI models. We start with the fixed-base model as the reference case, and progress to sophisticated models in terms of representing SSI phenomena. The first level is the lumped-parameter system using springs, point masses and dashpots, which is implemented in the time domain. At the highest level is the 3D FE model, and in between, the 2D FE model. A crucial question that arises is the effects of absorbing boundaries (Lysmer and Kuhlemayer 1969), which are used to truncate the semi-infinite FE mesh by blocking the reflection of outgoing elastic waves from the boundaries back into the core region. All these SSI models are then validated using various structural response indicators.

13.2   Modeling Dynamic SSI in a Conventional R/C Building

We begin with a three-story, moment-resisting reinforce concrete (R/C) frame and construct an equivalent single-degree-of-freedom (SDOF) oscillator as first model, with a mass of 17.63 t and a fundamental period of 0.5 s for fixed-base conditions. This SDOF system is subjected to a time-harmonic base acceleration and its inelastic time history response is recovered using the time stepping algorithm of Hilber-Hughes-Taylor (Hilber et al. 1977). More specifically, the SDOF system has a displacement ductility ratio $\mu = 2$ and a strength ratio $\eta = 1.5$. The supporting rectangular surface foundation has a length $L_f = 6$ m, width $B_f = 4$ m, modulus of elasticity $E_f = 3 \times 10^7$ kN/m^2 and rests on the homogeneous elastic half-space. The foundation is first assumed to be rigid and is then coupled to the impedance functions given by the Wolf (1994) and Gazetas (1991) models.

In order to reduce the computational effort, a 2D FE model with 27,786 DOF is introduced as an alternative to the fully 3D model with 168,044 DOF, see Fig. 13.1. The nodes in the 2D FE model have two in-plane DOF, namely the vertical and horizontal translations, while the out-of-plane DOF are suppressed. However, the 2D FE model used here is not a mere simplification of the 3D model, because the mesh has a large lateral extent in order to prevent the reflection of outward moving elastic waves. Consequently, no absorbing boundaries are applied to the 2D FE model outer boundaries. Finally, Table 13.1 lists numerical values for the soil mass density ρ, the Poisson's ratio ν, the shear-wave velocity c_s, the shear modulus G and the soil damping ratio ξ. For computational efficiency, the size of a soil FE is limited to 20 m or greater.

Fig. 13.1 The 2D (*top*) and 3D FEM (*bottom*) mesh used in the SSI modeling

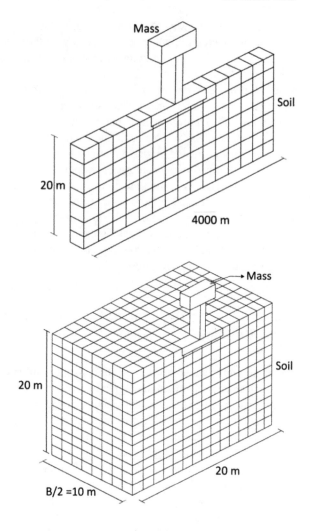

Table 13.1 Input values for the soil material parameters

ρ (kg/m^3)	ν	c_s (m/s)	G (N/m^2)	D (%)
1800	0.2	100	18×10^6	5

13.2.1 Hysteretic Material Models

The Takeda (Takeda et al. 1970) plus the Bouc-Wen (Wen 1976) hysteretic models are introduced in the inelastic time history analysis of the fixed-base structure, as well as for the superstructure coupled to the different soil models. More specifically, the Takeda model is one of the most commonly used in the nonlinear analysis of R/C structures and consists of sixteen rules for describing the tri-linear hysteresis loop. These rules, which are based on experimental data, govern the material stiffness characteristics at successive cycles of unloading, reloading, cracking and yielding. The Bouc-Wen model is represented by a nonlinear differential equation, which describes the restoring force behavior completely without additional conditions.

13.2.2   Modeling SSI Effects

Four different SSI models are used, namely the two simpler discrete parameter models of Wolf (1994) and Gazetas (1991), plus the two more complex FE models previously described, where the superstructure model is coupled to the FE model of the soil. The structural response is investigated for a horizontal base acceleration with harmonic (i.e., sinusoidal) time variation of amplitude 2 m/s^2 and period 0.5 s. This motion is applied at the base nodes of both 2D and 3D FE models, and travels upwards to the free surface as horizontally polarized shear (SH) waves. In the simpler Wolf and Gazetas models, the soil is replaced by spring, dashpot and added mass elements. This necessitates the computation of the free field motion from the original harmonic signal placed at bedrock and convoluted upwards to the soil-structure interface in the FE models (Elgamal et al. 2010). This, by definition, is the input motion that must be applied at the base of the superstructure and is amplified with respect to the original signal by a factor of almost three. Also, the structural response is also investigated without incorporating SSI effects by simply applying the free-field ground motion directly to the fixed base of the superstruc-ture. The foundation motion is not computed in the simpler Wolf and Gazetas models, which implies that kinematic interaction effects are not taken into account. Ignoring kinematic interaction is considered as a type of uncertainty in the numerical modeling process and is therefore characterized as an attribute of model complexity, as discussed in Sect. 13.4.

13.2.3   Top Story Displacement as Indicator

The results of the modal analysis performed for each of the five SSI models are summarized in Table 13.2. The results show the expected natural period elongation of the superstructure due to the flexibility of the supporting soil. Next, the relative displacement time histories at the top floor of the nonlinear, fixed-base structure, as well as of all four coupled SSI models under harmonic ground motion are computed and plotted in Fig. 13.2 for both the Takeda and Bouc-Wen hysteretic models. The

Table 13.2 Natural periods of the structure coupled to the soil models

Model	Fixed	Wolf	Gazetas	2D FE	3D FE
Period (s)	0.500	0.622	0.588	0.800	0.765

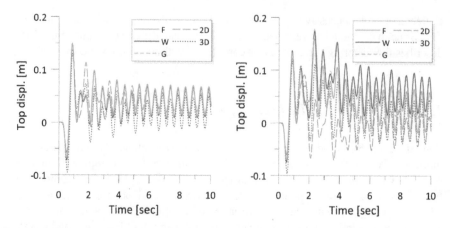

Fig. 13.2 Relative displacement time histories for SSI models subjected to harmonic ground motion (f = 2 Hz) using the Takeda (*left*) and the Bouc-Wen hysteretic models (*right*)

early time response, where all models behave linearly, is quite similar for both hysteretic models. Once the structure reaches the inelastic stage, its hysteresis loops become active and different maximum values are recorded in the displacements. More specifically, the fixed-base model plus the two simpler SSI models all give higher response values as compared to the two FE models. This can be explained by recourse to the natural periods presented in Table 13.2. Since the input motion frequency was set equal to the natural frequency of the fixed-base model, it is closer in value to the natural periods of the two simpler models.

13.3 Energy and Damage Measures for Dynamic Response

Housner (1956) initiated the use of the energy approach in seismic design and showed that energy is transmitted from the ground motions into a structure in various ways. Moreover, he pointed out that a part of this transmitted energy is absorbed within a structure in terms of recoverable elastic strain energy, while the remaining part is irrecoverable hysteretic energy when the response becomes inelastic. In here, we follow the relative energy approach proposed by Uang and Bertero (1988, 1990), where the response of the damped SDOF system subjected to a horizontal ground motion is expressed by the following equation of motion:

$$m\ddot{u} + c\dot{u} + r = -m\ddot{u}_g \tag{13.1}$$

In the above, m is the system's mass, u is the relative displacement of this mass with respect to the ground, $u_t = u_g + u$ is the total displacement, u_g is the ground displacement, c is the viscous damping coefficient and r is the restoring force. For linear elastic systems, $r = k\,u$, where k represents the stiffness. The viscous damping coefficient c is assumed to be constant, even as the system behaves nonlinearly. Integrating Eq. 13.1 with respect to u gives

$$\int m\ddot{u}du + \int c\dot{u}du + \int rdu = -\int m\ddot{u}_g du \qquad (13.2)$$

The above relative energy equation can be summarized as

$$E_k + E_d + E_a = E_i \qquad (13.3)$$

where the first term E_k refers to the relative kinetic energy and can be rewritten as

$$E_k = \int m\ddot{u}du = dt \int m(d\dot{u}/dt)du = \int md\dot{u}(\dot{u}) = m(\dot{u})^2/2 \qquad (13.4)$$

The damping energy term is energy dissipated by viscous damping up to time t:

$$E_d = \int c\dot{u}du = \int c\dot{u}^2 dt \qquad (13.5)$$

The third term is the absorbed energy, comprising a recoverable elastic strain energy E_s and an irrecoverable hysteretic energy E_h. Thus,

$$E_a = E_s + E_h = \int rdu, \text{ where} \qquad (13.6)$$

$$E_s = r^2/2k \qquad (13.7)$$

Finally, the E_h energy term materializes once the system responds nonlinearly and is computed as the sum of the areas delimited by each hysteresis loop traced in the force-displacement plot. Finally, the term on the right hand side of Eq. 13.2 is defined as the relative input energy and represents the work done by an equivalent static lateral force $\left(-m\ddot{u}_g\right)$ on the fixed base SDOF system, i.e.,

$$E_i = -\int m\ddot{u}_g du \qquad (13.8)$$

13.3.1 Energy Development During Inelastic System Response

The SDOF equation of motion can be normalized as shown below

$$\ddot{u} + 2\omega\xi\dot{u} + \omega^2 u = -\ddot{u}_g \tag{13.9}$$

where ω is the natural frequency of the system and ξ is the viscous damping ratio. Next, for an inelastic system response, Eq. 13.9 can be rewritten in non-dimensional form (Mahin and Lin 1983) as

$$\ddot{\mu} + 2\omega\xi\dot{u} + \omega^2 u = -\omega^2 \ddot{u}_g / \eta\ddot{u}_{g,\max} \tag{13.10}$$

where μ is the ductility ratio and η is the structural strength ratio, defined as

$$\mu = u/\delta_y, \quad \eta = r_y/m\ddot{u}_{g,\max} \tag{13.11}$$

In the above, r_y and δ_y are yield strength and yield displacement of the inelastic structure, respectively.

13.3.2 Energy Response to Harmonic Excitation

The dynamic inelastic response to a harmonic base acceleration of amplitude 2 m/s^2 and excitation frequency 2 Hz is investigated in this subsection. More specifically, the resulting energy response time histories (i.e., input energy, strain energy, kinetic energy and hysteretic energy) for the fixed-base structure, as well as for the coupled SSI models are all presented in Fig. 13.3 using the Bouc-Wen hysteresis material model for the superstructure. We observe that the cyclic inelastic structural response caused by the harmonic ground acceleration becomes more stable after the first few cycles of excitation. It can further be observed that the kinetic and strain energies fluctuate between the unbounded hysteretic energy and the input energy levels, while the resulting input energy keeps increasing as the hysteretic energy accumulates during the motion. Strain energy is at a maximum when the structure reaches its maximum displacement value, while kinetic energy is zero at that same instant. Also, strain energy is zero when the structure crosses the next static equilibrium point, with kinetic energy concurrently reaching a maximum value.

A comparison of the response of the coupled SSI systems to the response of the fixed-base structure shows that the former's energy response has decreased. However, the resulting energies produced by the structure when coupled to the 3D FE soil model in is relatively large in comparison to the response of all other models. A probable reason for this higher response is the use of Lysmer-type absorbing boundaries along the vertical borders of the 3D soil continuum. These viscous boundaries simulate the transmission of wave energy from the core region to the outer half-plane. However, full energy absorption of reflected waves cannot be assured (Kramer 1996) since these boundaries can only absorb elastic waves that strike the boundary at an angle of incidence less than 90° with respect to the

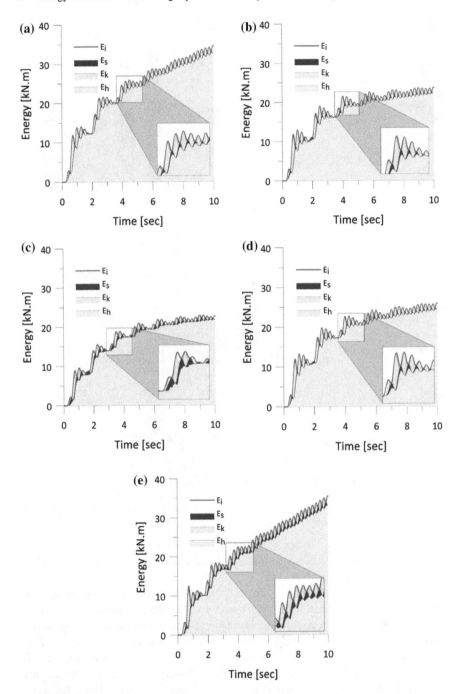

Fig. 13.3 Energy time histories for a structure with Bouc-Wen hysteresis under harmonic ground motion (f = 2 Hz): **a** Fixed-base; **b** Wolf; **c** Gazetas; FE 2D; **e** FE 3D models. *Legend* E_i = input energy; E_s = strain energy; E_k = kinetic energy; E_h = hysteretic energy

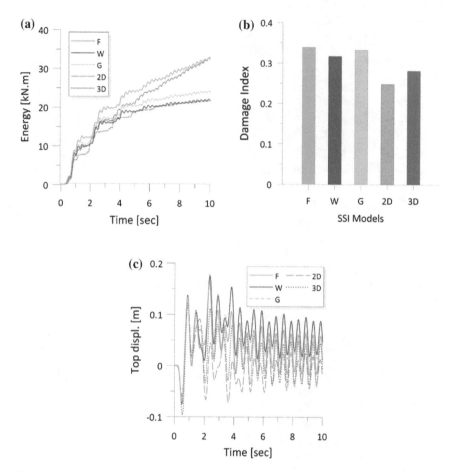

Fig. 13.4 Structural **a** hysteretic energy, **b** damage index and **c** structural top story displacement for models with Bouc-Wen hysteresis and subjected to harmonic ground motion (f = 2 Hz). *Legend* F = Fixed-base; W = Wolf model; G = Gazetas model; 2D = 2D FEM model; 3D = 3D FEM model

horizontal plane. In our case, these angles of incidence on the vertical boundaries of the FE models vary from 0° to 180°.

Next, Fig. 13.4 depicts the structural responses computed by using the Bouc-Wen hysteretic model. This response is given by three indicators for each of the five models studied, i.e., the time history response of the cumulative hysteretic energy dissipated in a structure (an indicator of the damage sustained), the time history response of the top story displacement, and the Park and Ang (1985) damage index. The displacement response time histories for the different models are for an undamped system, which implies zero damping energy. Thus, the input energy is the sum of kinetic, hysteretic and strain energies and the energy plots for the different models satisfy the principle of energy balance for the structure under consideration. More specifically, Fig. 13.4 shows that the fixed-base model dissipates the largest amount of hysteretic energy in comparison to all other models investigated. An exact correspondence between the order of the models regarding damage on one hand, and the amount of absorbed hysteretic energy on the other hand, can be observed in the time range of 0–2 s. More often than not, the response of the 2D FE model delineates the lower bound of the energy response. A relation between energy demand and damage sustained is clearly observed, where the highest magnitude of energy response produced by the fixed-base structure is associated with the highest damage grade. Conversely, the lowest damage grade is observed in the 2D FE model, which dissipates the lowest amount of hysteretic energy.

All previous observations also apply for the estimates (not presented here in the interest of brevity) for the Takeda hysteresis model is used. However, the grade of damage sustained by the different models is lower in comparison to the corresponding results when the Bouc-Wen hysteresis model is used. This can be explained by the fact that the smaller amount of hysteretic energy dissipation is manifested in the Takeda model.

13.4 Model Quality Assessment

Following EPA (2009) guidelines, uncertainty refers to incomplete knowledge about specific factors, parameters, input, or models. More specifically, we have:

Model framework uncertainty: This type of uncertainty is produced by lack of knowledge regarding the theoretical background of the structural system modeled, i.e., paucity in identifying and assigning values to the factors that influence the true behavior of the system and the possible idealizations and simplifications of the original system.

Model input uncertainty: This type of uncertainty is produced by data measurement errors, an inadequate amount of sample input data and by the use of stochastic description of the system parameters involved, stemming from the model's natural variability and inherent randomness.

On the other hand, *model complexity* is considered to be a major factor that influences the quality of a numerical model. Increased model complexity usually means that more parameters are required to describe it. Consequently, more input data is needed, which must be obtained either through field measurements or estimated empirically. Also, input parameters require initial conditions that are defined by the underlying modeling assumptions. In sum, Fig. 13.5 above illustrates the relationship between the different types of uncertainty, indicating how the degree of complexity is associated with total model uncertainty.

Fig. 13.5 Relationship
between model framework
uncertainty and input
parameters uncertainty and
the resulting total model
uncertainty adapted after EPA
(2009)

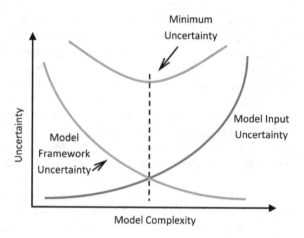

13.4.1   The Adjustment Factor Approach

Following Most (2011) and Park et al. (2010), the adjustment factor approach is
used to predict a system response Y_{pred} from a set of models as follows:

$$Y_{pred} = y^* + E_a^*$$ (13.12)

In here, y^* represents a prediction of the response, as produced by the reference
model. The latter is first adopted as a more complex model, which provides a
detailed representation of the real system since it uses fewer idealizations and
simplifications compared to the averaged model. Finally, E_a^* is an additive adjustment factor that represents the error in the model response.

By including uncertainty in the model parameters, the mean and variance of the
predicted response Y_{pred} are computed as follows:

$$E\left(Y_{pred}\right) = \bar{Y}^M = \sum_{i=1}^{k} P_{M_i} E\left(Y^{M_i}\right)$$ (13.13)

$$V\left(Y_{pred}\right) = V\left(Y^M\right) = \sum_{i=1}^{k} P_{M_i} E\left(Y^{M_i} - \bar{Y}^M\right)^2$$ (13.14)

In the above, Y^{M_i} represents the reference system response, P_{M_i} is the probability
associated with model M_i, and k is the number of available models. The model
probability P_{M_i} can be assumed equal to $(1/k)$ by considering a uniformly distributed, discrete variable model. Alternatively, P_{M_i} can be considered as a
weighting factor, with the different values adding up to unity. The modified model
response $Y^{M_i^*}$ introduced by Most (2011) is based on an adopted reference response
with additive model errors, which is represented as

$$Y^{M_i^*} \approx Y^{M_i} + \varepsilon_\Delta^{M_i} + \varepsilon^{M_{ref}} \tag{13.15}$$

In the above, $\varepsilon_\Delta^{M_i}$ denotes the model framework uncertainty as an additive error to the reference model response, while $\varepsilon^{M_{ref}}$ is the error associated with the reference model response. The variance of $\varepsilon_\Delta^{M_i}$ is then approximated as

$$V\left(\varepsilon_\Delta^{M_i}\right) \approx b^2 \left(\bar{Y}^{M_i} - \bar{Y}^{M_{ref}}\right)^2 \tag{13.16}$$

where $\bar{Y}^{M_{ref}}$ is the reference response, and $b = 0.608$, as corresponding to a 95% one-sided quantile. The total variance of a numerical model under consideration can then be written as

$$V\left(Y^{M_i^*}\right) \approx V\left(Y^{M_i}\right) + b^2 \left(\bar{Y}^{M_i} - \bar{Y}^{M_{ref}}\right)^2 + V\left(\varepsilon^{M_{ref}}\right) \tag{13.17}$$

In conclusion, Eq. 13.17 indicates that the best model from a set of possible ones is the one with minimum total variance, i.e., the smallest sum of model input uncertainty and model framework uncertainty.

13.4.2 Numerical Results

Four structural response indicators are used to estimate uncertainty in the SSI models investigated here, namely the maximum top story displacement of the structure d_{max}, the ratio of total structural hysteretic energy to total structural input energy E_h/E_i, the Park-Ang damage index DI_{PA} and the averaged structural top story displacement d along a predetermined time window, truncated from the entire response time history. Two types of hysteretic rules are used (Takeda and Bouc-Wen) in the nonlinear analysis, while the SSI models are subjected to two different types of excitation, namely the harmonic base acceleration previously described and a group of six Ricker wavelet pulses Ryan (1994) of variable frequency content ranging from 1.0–2.0 Hz. The wavelet acceleration amplitude is equal to $a_{max} = 2$ m/s^2.

13.4.3 Uncertainty in the Model Framework

In this subsection, uncertainty in the model framework is investigated for all SSI models, while at the same time, effects due to uncertainty in the input parameters are ignored. This uncertainty for each SSI model is estimated based on a reference response, which is the averaged model response derived from the responses of all individual models. Different probabilities (weights) are assigned to the five SSI models, since they have different abstraction levels. After the fixed-base model, the Gazetas and Wolf models are considered to be the simpler SSI models since the surrounding soil continuum is replaced with a spring-dashpot-mass system. Each of these three models has a weighted model prediction equal 1, which corresponds to a probability that is equal to 1/9 of the sum of all model probabilities. The 2D and 3D FEM models are more complex, and as such are more representative of the real system than the simpler models. Therefore, each is given a larger weighted prediction equal to 3, which corresponds to a probability equal to 3/9 of the sum of all model probabilities. These values are subsequently applied as weights P_{Mi} in

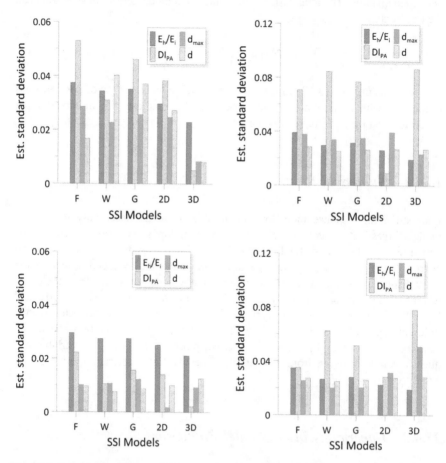

Fig. 13.6 Estimated *model framework uncertainty* for models subjected to (*top*) harmonic ground motion (f = 2 Hz) and (*bottom*) pulse ground excitation (f = 1.0 Hz) with averaged response as reference: (*left*) Bouc-Wen (*left*) and (*right*) Takeda hysteresis models. *Legend* F = Fixed-base; W = Wolf model; G = Gazetas model; 2D = 2D FE model; 3D = 3D FE model; $E_h = E_i$ = ratio of total hysteretic energy to total input energy; DI_{PA} = Park-Ang damage index; d_{max} = maximum top story displacement; d = averaged top story displacement

Eqs. 13.13 and 13.14. Next, Fig. 13.6 plots the deterministic predictions given by the four SSI models plus the fixed-base model for estimated *model framework uncertainty*. An overview of these results, which represent model-to-model uncertainty with the averaged response as reference, can be found in Table 13.3 to provide better perspective. In general, the response indicators lead to a smaller model error when using the Takeda as compared to the Bouc-Wen hysteresis model. The results show that the hypothesis of decreasing model error with increasing model complexity is empirically validated. At the same time, ignoring the uncertainty of the input parameters, holds true only when quality assessment is based on energy ratio E_h/E_i as the response indicator. An estimation of model uncertainty based on E_h/E_i and on damage index DI_{PA} leads to models with the same order of estimated quality, as shown in Fig. 13.6. This conclusion holds true in reference to the aforementioned results only when based on the energy ratio as the response indicator.

In sum, these results basically apply for both types of hysteretic models and show that the more complex, 3D FE model has the best quality of all models investigated. When one focuses on the three simpler models, the Wolf model produces the lowest model uncertainty, and therefore exhibits the best model quality compared to the other two. Finally, the fixed-base model produces the highest estimated uncertainty and consequently the worst quality of all models investigated. Once again, this confirms the hypothesis of increasing model error with decreasing complexity, but only when this assessment is based on the energy ratio used as a response indicator. The results in Fig. 13.6 show that the estimated model framework uncertainty is dependent on the type of loading function and on the excitation frequency (note that results are shown for one pulse frequency only in the interest of brevity). However, the estimation based on the energy ratio E_h/E_i is frequency-independent, and at the same time this ratio leads to a less sensitive estimation of model quality.

13.4.4   Total Model Uncertainty

In addition to the uncertainty in the model framework, uncertainty in the SSI model input parameters is also estimated in this subsection. More specifically, uncertainty in the model parameters is investigated by means of the Latin hypercube sampling method (McKay et al. 1979). This sampling is performed by setting up independent log-normal distributions with mean values μ and a coefficients of variation cov, see Table 13.4. Three parameters selected as stochastic in this analysis, namely the soil shear wave velocity c_s, the structural strength ratio η and the foundation modulus of elasticity E_f. Eight samples are randomly generated (EPA 1997) for the soil and the structural variables. The objective is to measure the relative effect of an uncertain input parameter on the total uncertainty of a model, so long as this input parameter is not correlated with parameters used in the remaining models. In addition to the averaged model response used above, the response of the 2D FE model is also used

Table 13.3 Model quality predictions corresponding to the *model framework uncertainty* illustrated in Fig. 13.6

Indicator	f (Hz)	Fixed	Wolf	Gazetas	2D FE	3D FE
E_h/E_i	1.0	E	C	D	B	A
	1.2	E	C	D	B	A
	1.4	E	C	D	B	A
	1.6	E	C	D	B	A
DI_{PA}	1.0	B	D	C	A	E
	1.2	B	D	C	A	E
	1.4	A	D	C	B	E
	1.6	A	D	B	C	E
d_{max}	1.0	D	B	C	E	A
	1.2	C	A	B	D	E
	1.4	C	B	A	D	E
	1.6	C	B	A	D	E
d	1.0	E	A	B	D	C
	1.2	D	A	B	C	E
	1.4	D	A	B	C	E
	1.6	D	A	C	B	E

Legend A = best quality; E = worst quality

Table 13.4 Stochastic soil and structural variables for a log-normal distribution

c_s (m/s)		η		E_f (N/m^2)	
μ	cv	μ	cv	μ	cv
175	0.4	1.5	0.15	29.9×10^9	0.03

as a reference model. Again, the results presented here represent model-to-model uncertainty. The reason to adopt the 2D FE model as reference is because it is considered to give the best real system approximation of all models investigated. The soil medium in this FE model has a width of 4 km, which is large enough to prevent the reflection of the waves from the vertical model boundaries. This width was computed by considering the duration of the input excitation and the speed of the elastic waves travelling through soil. Thus, there is no need for implementing artificial absorbing boundaries in this model.

Accounting therefore for material parameter uncertainty in the quality estimation of the SSI models changes the order of models regarding their total uncertainty. The fixed-base model, which only has the structural strength ratio η as the sole stochastic input parameter, has the lowest total error as shown in Fig. 13.7. These results show quantitatively how the uncertain input parameters affect total estimated uncertainty, with the differences in the estimated uncertainties become smaller between the simpler model group and the more complex model group after including uncertainty in the material parameters. On one hand, the three simpler models have fewer input parameters and are therefore less sensitive to the uncertainty associated with them. On the other hand, the two more complex FE models show a significant increase in the total error, once the uncertainty of their input

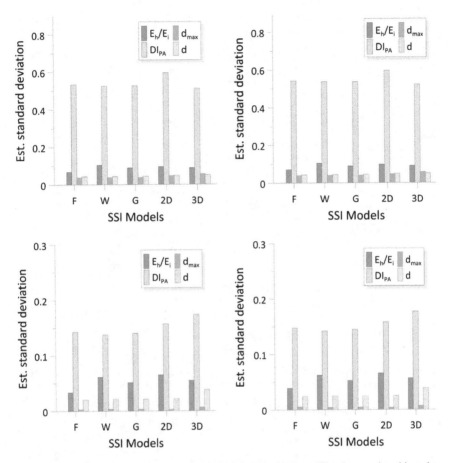

Fig. 13.7 Estimated *total model uncertainty* for models with Bouc-Wen hysteresis subjected to (*top*) harmonic ground excitation (f = 2 Hz) and (*bottom*) pulse ground excitation (f = 1 Hz): (*left*) the 2D FE model and (*right*) the averaged model response as a reference. *Legend* F = Fixed-base; W = Wolf model; G = Gazetas model; 2D = 2D FE model; 3D = 3D FE model; $E_h = E_i$ = ratio of total hysteretic energy to total input energy; DI_{PA} = Park-Ang damage index; d_{max} = maximum top story displacement; d = averaged top story displacement

parameters is included. To put it simply, the FE models use more parameters, and consequently produce a higher total uncertainty compared to the three simpler SSI models. Finally, although Fig. 13.7 show the estimation results for the SSI models subjected to one pulse for brevity, Tables 13.5 and 13.6 providing a summary of results for four pulse frequencies ranging from 1.0–1.6 Hz. Once again, we observe that an estimation based on the energy ratio E_h/E_i as the response indicator leads to the same order of classification of the SSI models in terms of their quality. This also holds true for an estimation based on both types of reference response (i.e., 2D FE and averaged) shown in Fig. 13.7, but this observation, however, does not apply for the remaining three response indicators.

Table 13.5 Model quality predictions corresponding to the *total model uncertainty* illustrated in Fig. 13.7 with 2D FE as the reference response

Indicator	Freq. (Hz)	Fixed-base	Wolf	Gazetas	2D FE	3D FE
E_h/E_i	1.0	A	D	B	E	C
	1.2	A	D	B	E	C
	1.4	A	D	B	E	C
	1.6	A	D	B	E	C
DI_{PA}	1.0	C	A	B	D	E
	1.2	D	A	B	C	E
	1.4	D	A	C	B	E
	1.6	D	B	C	A	E
d_{max}	1.0	B	D	C	A	E
	1.2	B	D	C	A	E
	1.4	B	D	C	A	E
	1.6	B	C	D	A	E
d	1.0	A	C	B	D	E
	1.2	A	D	C	B	E
	1.4	B	D	C	A	E
	1.6	A	D	B	C	E

Legend A = best quality; E = worst quality

Table 13.6 Model quality predictions corresponding to the *total model uncertainty* illustrated in Fig. 13.7 with averaged model as reference response

Indicator	Freq. (Hz)	Fixed-base	Wolf	Gazetas	2D FE	3D FE
E_h/E_i	1.0	A	D	B	E	C
	1.2	A	D	B	E	C
	1.4	A	D	B	E	C
	1.6	A	D	B	E	C
DI_{PA}	1.0	C	A	B	D	E
	1.2	D	A	B	C	E
	1.4	D	B	C	A	E
	1.6	D	B	C	A	E
d_{max}	1.0	D	B	C	A	E
	1.2	D	C	B	A	E
	1.4	D	B	C	A	E
	1.6	D	B	C	A	E
d	1.0	A	C	B	D	E
	1.2	A	D	C	B	E
	1.4	A	D	B	C	E
	1.6	A	C	B	D	E

Legend A = best quality; E = worst quality

13.5   Conclusions and Discussion

The methodology developed here for evaluating the quality of dynamic SSI models employs energy measures, which a physically robust concept. As a result, it provides the means to help structural engineers in selecting appropriate numerical models, despite all the uncertainties associated with SSI. More specifically, results show a certain degree of methodological independence from the frequency content of the input base excitation, as well as from the choice of structural hysteresis rules. This produces quite general results in terms of their applicability. Given the degree of uncertainty in the input material parameters, engineers can make informed decisions on the selection of one particular SSI model over another. Thus, the amount of uncertainty which can be tolerated in a numerical model implementation can be decided upon. This helps reduce both effort and cost involved in numerical simulations by more complex models, since our case study shows that these models are not of better quality as compared to simpler ones. Significant SSI influence can be observed in the structural response when using more complex, truncated soil models. This can be explained by the ability of these models to incorporate more factors, such as kinematic interaction and foundation flexibility, in addition to the more realistic representation of the surrounding soil medium and of its stress-strain response. The coupled soil-structure models generally show a decrease in the energy response. However, the energy response that 3D FE models produce seems to be relatively high in comparison to that of other coupled models, which is due to spurious elastic wave reflections from the outer model boundaries.

The ratio of structural hysteretic energy to input energy E_h/E_i yields robust predictions for evaluating quality in SSI models, in contrast to predictions based on the maximum floor displacement response. The resulting energies thus exhibit this desirable independence between structural response and frequency content of the input signals, which is clearly demonstrated when pulse-type wavelets are used. Also, predictions based on energy ratio E_h/E_i are insensitive to the use of different hysteretic rules. In other words, the estimation of model uncertainty based on energy ratio E_h/E_i in all scenarios examined leads to the same ordering of the SSI models as regards to their quality. Further investigations on model framework uncertainty show that the more complex 3D FE model has the best quality of all models investigated, whereas the Wolf SSI model produces the lowest model uncertainty between the three simpler models. The fixed-base model produces the highest estimated uncertainty and accordingly the worst result quality of all models investigated.

Despite the good correlation between the Park-Ang damage index and the hysteretic energy dissipated in the superstructure for models producing upper and lower bounds of damage grades, it can be seen that the best model quality does not necessarily correspond to a conservative structural design resulting in the lowest damage grade. When accounting for uncertainty in the input material parameters during a quality estimation of the SSI models, the order of models regarding their total uncertainty changes. Also, differences in the estimated uncertainties become smaller between the simpler and the more complex models, after uncertainty in the input parameters is accounted for. The simpler models have fewer input parameters and are therefore less sensitive to uncertainty, while the more complex models produce an increase in the total error after including the input parameter uncertainty. Finally, the computed uncertainty in model response is directly related to the model predictions used, which in turn are considered as weights within the framework of the *adjustment factor* approach. Thus, misrepresented model predictions might affect the resulting uncertainty considerably. Future research should therefore focus on ways to incorporate the uncertainty in model predictions into the evaluation process. Also, the evaluation method using energy measures can be extended to solve other related types of problems in structural engineering, such as the efficiency of different base isolation systems.

References

Beck MB (2006) Model evaluation and performance. Encyclopedia of environmetrics. Wiley, New York

Elgamal A, Yang Z, Parra E, Ragheb A (2010) Cyclic 1D-Internet-based computer simulation of site earthquake response and liquefaction. University of California, San Diego. http://cyclic.ucsd.edu/references.html

EPA (1997) Guiding principles for Monte Carlo analysis. Report no EPA-630-R-97-001, Environmental Protection Agency, Washington, DC

EPA (2009) Guidance on the development, evaluation and application of environmental models. Report no EPA/100/K-09/003, Environmental Protection Agency, Washington, DC

Gazetas G (1991) Formulas and charts for impedances of surface and embedded foundations. J Geotech Eng ASCE 117(9):1363–1381

Hilber HM, Hughes TJR, Taylor RL (1977) Improved numerical dissipation for time integration algorithms in structural dynamics. Earthq Eng Struct Dyn 5:283–292

Housner GW (1956) Limit design of structures to resist earthquakes. In: Proceedings of 1st world conference on earthquake engineering, Berkeley, California, pp 5.1–5.13

Konikow LF, Bredehoeft JD (1992) Ground-water models cannot be validated. Adv Water Resour 15:75–83

Kramer SL (1996) Geotechnical earthquake engineering. Prentice-Hall, Englewood Cliffs, New Jersey

Luis SJ, McLaughlin D (1992) A stochastic approach to model validation. Adv Water Resour 15:15–32

Lysmer J, Kuhlemeyer R (1969) Finite dynamic model for infinite media. J Eng Mech Div ASCE 95:859–875

Mahin SA, Lin J (1983) Construction of inelastic response spectra for single-degree-of freedom systems. Technical report no EERC-83/17, Earthquake Engineering Research Centre, University of California, Berkeley

McKay MD, Beckman RJ, Conover WJ (1979) A comparison of three methods for selecting values of input variables in analysis of output from a computer code. Technometrics 21 (2):239–245

Most T (2011) Assessment of structural simulation models by estimating uncertainties due to model selection and model simplification. Comput Struct 89(17–18):1664–1672

Naiem F (2001) The seismic design handbook, 2nd edn. Kluwer Academic Publications, Dordrecht

Nasser M, Schwedler M, Wuttke F, Könke C (2010) Seismic analysis of structural response using simplified soil-structure interaction models. D-A-CH-Mitteilungsblatt, Swiss Soc Earthq Eng 85:10–16

Oreskes N, Shrader-Frechette K, Belitz K (1994) Verification, validation and confirmation of numerical models in the Earth sciences. Science 263:641–646

Pearl J (1978) On the connection between the complexity and credibility of inferred models. Int J Gen Syst 4:255–264

Rahnama M, Manuel L (1996) The effect of strong motion duration on seismic demands. In: Proceedings of 11th world conference on earthquake engineering, Acapulco, Mexico, paper no 108

Park I, Amarchinta HK, Grandhi RV (2010) A bayesian approach for quantification of model uncertainty. Reliab Eng Syst Saf 95:777–785

Park YJ, Ang AHS (1985) Mechanistic seismic damage model for reinforced concrete. J Struct Eng ASCE 111:722–739

Ryan H (1994) Ricker. A choice of wavelets, CSEG recorder, Ormsby, Klander, Butterworth, pp 8–9

Takeda T, Sozen MA, Nielsen NN (1970) Reinforced concrete response to simulated earthquakes. J Struct Eng ASCE 96(12):2557–2573

Uang CM, Bertero VV (1988) Use of energy as a design criterion in earthquake resistant design. Technical report EERC-88/18, Earthquake Engineering Research Centre, University of California, Berkeley

Uang CM, Bertero VV (1990) Evaluation of seismic energy in structures. Earthq Eng Struct Dyn 19:77–90

Wen YK (1976) Method for random vibration of hysteretic systems. J Eng Mech Div ASCE 102:249–263

Wolf JP (1994) Foundation vibration analysis using simple physical models. Prentice-Hall, Englewood Cliffs, NJ

Chapter 14
Numerical and Experimental Identification of Soil-Foundation-Bridge System Dynamic Characteristics

P. Faraonis, Frank Wuttke and Volkmar Zabel

Abstract The natural frequencies of the Metsovo bridge during construction are identified both in actual scale and in 1:100 scale. Finite element models of increasing modeling complexity are developed in order to investigate their efficiency in representing the measured dynamic stiffness of the bridge-foundation-soil system. The results highlight the importance of accurately simulating boundary conditions in Structural Health Monitoring applications.

Keywords Soil-structure interaction · System identification · Finite element modeling

P. Faraonis (✉)
Department of Civil Engineering, Aristotle University of Thessaloniki,
54124 Thessaloniki, Greece
e-mail: pfaraoni@civil.auth.gr

F. Wuttke
Chair of Marine and Land Geomechanics and Geotechnics,
Kiel University, Ludewig-Meyn St. 10, 24118 Kiel, Germany
e-mail: fw@gpi.uni-kiel.de

F. Wuttke
Institute of Applied Geo-Sciences, Christian-Albrecht's University,
24118 Kiel, Germany

F. Wuttke
Faculty Civil Engineering, Formerly, Bauhaus-University Weimar,
Coudraystrasse 11C, 99423 Weimar, Germany

V. Zabel
Research Associate and Lecturer, Bauhaus-University Weimar,
Marienstrasse 15, 99421 Weimar, Germany
e-mail: volkmar.zabel@uni-weimar.de

© Springer International Publishing AG 2017
A.G. Sextos and G.D. Manolis (eds.), *Dynamic Response of Infrastructure to Environmentally Induced Loads*, Lecture Notes in Civil Engineering 2,
DOI 10.1007/978-3-319-56136-3_14

14.1 Introduction

The dynamic characteristics of structures can be either identified through System
Identification (SI) methods or predicted by modal analysis of numerical finite
element (FE) models. System Identification methods can identify the modal properties of structures by measuring their response to a known excitation (input-output
methods, Wemer et al. 1987; Chaudhary et al. 2000) or to an unknown excitation
(output-only methods, Basseville et al. 2001; Peeters and De Roeck 2001). The
modeling assumptions of the FE models can be evaluated by comparing the
identified with the predicted modal characteristics. A wide variety of studies
(Crouse et al. 1987; Chaudhary et al. 2001 and Morassi and Tonon 2008) present
the influence of soil stiffness to the SI results and the importance of taking into
account soil compliance in FE models, in order to minimize the discrepancies
between identified and numerically predicted dynamic characteristics.

One option to account for soil compliance is by numerically modeling the entire
structure-foundation-soil system as a whole (Wolf 1989). Due to the fact that this
method is quite expensive from a computational standpoint and is not easily
implemented in engineering practice, alternative methods have also been developed. In these methods the structure-foundation-soil interaction is decoupled to
kinematic and inertial component. As far as the shallow embedded foundations are
concerned, it is common to replace the foundation-soil system with six
degrees-of-freedom (DOF) springs, the stiffness of which is calculated according to
Elsabee and Morray (1977). Alternatively, the subsoil may be replaced by 6-DOF
springs concentrated at the base of the foundation (defined according to Kausel
1974) as well as additional springs attached on the foundation (Wolf 1989).
Experimental and numerical evaluation of the efficiency of the aforementioned
methods in representing the dynamic stiffness of various foundation-soil systems is
presented by Varun and Gazetas (2009).

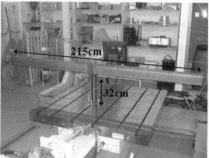

Fig. 14.1 Metsovo Bridge segments during the construction stage (*left*) and its equivalent scaled
structure tested at the laboratory (*right*)

In this framework, the scope of this paper is to experimentally verify the influence of soil compliance on the predictions of System Identification and to investigate the efficiency of existing numerical methods in simulating the soil stiffness.

14.2 Prototype Structure

The Metsovo ravine bridge was constructed in 2008 in Greece along the Egnatia Highway and consists of two structurally independent branches (one for each carriageway). The bridge was constructed by the balanced cantilever construction method, which made feasible the modal identification of structurally independent bridge components during construction. The modal characteristics of the M3 (cantilever) pier (Fig. 14.1, left) were identified prior to the construction of the key connecting segments to the M2 pier, the latter also temporary acting as a balanced cantilever (Panetsos et al. 2009). The modal identification of the M3 cantilever was based on ambient vibration measurements triggered by wind and induced operational loads. Detailed information regarding the measurements and the applied identification methodology can be found in Panetsos et al. (2009).

14.3 Scaled Structure with Alternative Boundary Conditions

A scaled model structure of the prototype M3 pier cantilever of Metsovo Bridge was constructed in the laboratory of Soil Mechanics at the Bauhaus University Weimar. Apart from the stiff foundation soil corresponding to the actual conditions of the prototype structure, alternative boundary conditions were also examined in the form of gradually stabilized soil to investigate the influence of soil compliance on the prediction of modal characteristics.

14.3.1 Scaled Structure Fixed

The construction of a scaled structure dictates the determination of the scaling laws relating the prototype geometry to that of the scaled structure. The scaling laws can be determined either by dimensional analysis or the analysis of the system's characteristic equation. Based on dimensional analysis and by neglecting the gravity distortion effects that arise during scaling, the scaling factor that relates the natural frequencies of a scaled structure with its prototype is given in Eq. (14.1) (Bridgman 1931):

Fig. 14.2 Scaled structure on stabilized soil

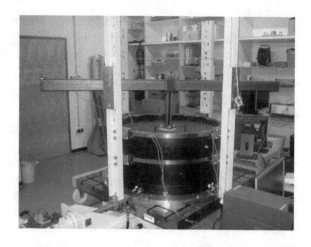

$$\lambda_f = \frac{1}{\lambda_l} \cdot \sqrt{\frac{\lambda_E}{\lambda_\rho}} \qquad (14.1)$$

where

λ_f is the prototype to the model frequency ratio,
λ_l is the prototype to the model dimension ratio,
λ_E is prototype to the model young modulus of Elasticity ratio,
λ_ρ is prototype to the model density ratio

Herein, the scale set to 1:100, to accommodate the fact that the deck length of the prototype structure is 215 m (Fig. 14.1). It is noted that as the exact section of the concrete deck could not be reproduced at a 1:100 scale (i.e., the resulting web and flanges would be as thin as 22 and 3 mm) an equivalent steel structure with the same dynamic characteristics was formed in the laboratory based on parametric modal analysis. Several standard steel sections were considered until matching with the modal characteristics of the concrete scaled structure was achieved. The equivalent, steel balanced cantilever was finally formed by the following commercially available sections:

a 90X90X3 HSS steel hollow section of 215 cm length corresponding to an 1:100 replication of the prototype deck,
a 100X100X5 HSS steel hollow section of 6.15 cm length corresponding to the prototype central deck-segment, and
a 80X20X3 HSS steel hollow section of 32 cm length corresponding to 1:100 replication of the prototype M3 pier.

14.3.2   Scaled Structure on Stabilized Soil

Next, the scaled structure was fixed on a circular concrete foundation (15 cm diameter and height). The structure was then placed within a 95 cm diameter laboratory box that was filled with stabilized soil (Fig. 14.2). The stabilized soil consisted of clay (CL), with 24% water and 4% lime. The latter was added in order to increase the stiffness of the soil and its percentage was determined according to DIN EN 459-1. The total height of the stabilized soil in the box was 30 cm and the dry density was determined as $\rho_s = 1.86$ t/m^3. Sensors were placed inside the box in order to measure the shear wave velocity of the stabilized soil.

14.4   Finite Element Models

14.4.1   Fixed Conditions

A refined finite element (FE) model using three-dimensional solid elements was developed to simulate the fixed scaled structure as shown in Fig. 14.1 (right). The resulted FE model consisted of approximately 19,000 triangular elements corresponding to 88,620 degrees of freedom. The measured mass of the physical model was 20.46 kg with a density of $\rho = 7.46$ t/m^3. The modulus of elasticity of the stainless steel was taken equal to 210 GPa.

14.4.2   Soil Compliant Conditions

Three methods were used to simulate the dynamic stiffness of the soil-foundation system of the scaled structure that was founded on stabilized soil (Fig. 14.2).

In the first method, a three-dimensional finite element model of approximately 200,000 degrees of freedom was further developed for the entire pier-foundation-subsoil system. Stainless steel (E = 210 GPa, v = 0.3) was once more assigned to the superstructure, whereas C30/37 concrete properties (E = 32 GPa, v = 0.3) were assigned to the caisson. The mass of the foundation was measured 7.56 kg corresponding to a density $\rho = 2.71$ t/m^3. The shear modulus of the stabilized soil was taken equal to G = 186 MPa based on the shear wave velocity $V_S = 316$ m/s and density $\rho_s = 1.86$ t/m^3 values that were measured in the laboratory.

In the second method, soil was modeled through 6-DOF springs at the base of the foundation and by 6-DOF springs at the middle of the foundation height. The stiffness of the former springs was obtained from the theory of rigid circular foundations on a stratum over rigid base suggested by Kausel (1974) while the stiffness of the latter springs was calculated by the solution of Varun and Gazetas

Table 14.1 Identified and numerically predicted natural frequencies for the case that the scaled structure was fixed at its base

Modes	Fixed boundary conditions			
	Prototype		Equivalent scaled structure	
	Identified	Expected	Identified	Numerical model fixed
		Ideal 1:100 (not constructed)		
	(Hz)	(Hz)	(Hz)	(Hz)
Rotational	0.159	15.90	15.96	16.01
1st longitudinal	0.305	30.50	23.67	23.13
Transverse	0.623	62.30	65.56	68.23
2nd longitudinal	0.686	68.60	67.68	69.71
Bending (deck)	0.908	90.80	88.65	89.45
Average Δf (%)			**6.34**	**2.12**

(2009) for cylindrically shaped large diameter caisson foundations. In these formulas the soil shear modulus G was again estimated based on the measured V_S of the stabilized soil.

In the third method, both the foundation and the soil were replaced by 6-DOF Winkler type springs. Their values were obtained from the theory of rigid embedded cylindrical foundations welded into a homogenous soil stratum over bedrock, proposed by Elsabee and Morray (1977). Again the G shear modulus was also estimated based on the measured V_S of the stabilized soil.

14.5 Results

14.5.1 Prototype Structure Versus Equivalent Scaled Structure

The first five identified natural frequencies of the M3 cantilever prototype structure range between 0.159–0.908 Hz and are presented in Table 14.1. The corresponding natural frequencies that are theoretically anticipated for an 1:100 scaled structure, ideally comprising of the same material, are also presented in Table 14.1 and vary between 15.90 and 90.80 Hz. These expected natural frequencies are used to validate the equivalence of the constructed scaled (steel) structure with the prototype.

The equivalent scaled (steel) structure was subjected to hammer impulses in order to simulate a broad band excitation, similar to ambient excitations applied to the actual M3 cantilever. The natural frequencies were identified by the stochastic subspace identification method (Peeters and De Roeck 2001) with the use of

Table 14.2 Identified and numerically predicted natural frequencies for the case that the foundation of the scaled structure was embedded in stabilized soil

Modes	Stabilized soil boundary conditions			
	Identified	Numerical models		
	Stochastic subspace identification (Hz)	Method 1 Holistic method (Hz)	Method 2 6 + 6 DOFs springs (Hz)	Method 3 6 DOFs springs (Hz)
Rotational	14.88	15.63	15.64	15.82
1st longitudinal	19.15	21.74	21.69	21.98
Transverse	46.52	57.42	56.77	60.6
2nd longitudinal	56.87	63.80	63.34	66.59
Bending (deck)	85.25	88.77	88.72	88.76
Average Δf (%)		**11.66**	**11.16**	**14.51**

MACEC, which is a Matlab toolbox for operational modal identification. The first five identified natural frequencies given in Table 14.1 range between 15.96–88.65 Hz. It has been observed that the natural frequencies of the equivalent scaled structure present a 6.34% average deviation compared to those expected from the prototype's ideal 1:100 scaled structure, indicating good agreement between the equivalent (steel) and the prototype (concrete) bridge pier.

14.5.2 Identified and Numerically Predicted Natural Frequencies

14.5.2.1 Fixed Boundary Conditions

Next, it is verified that the first five natural frequencies predicted by the fixed FE model and range between 16.01–89.45 Hz are in very good agreement with those of the tested equivalent structure showing only a 2.12% average error as are summarized in Table 14.1.

14.5.2.2 Stabilized Soil as Foundation Soil

A hammer impulse excitation was also applied to identify the natural frequencies of the scaled structure when the latter was placed within the stabilized soil. The first five identified natural frequencies are presented in Table 14.2, and range between 14.88–85.25 Hz. In Table 14.2 it is also observed that the average deviation between the identified and the numerical predicted frequencies range is of the order of 11–14%. Given that the experimentally and numerically predicted natural frequencies of the fixed system were almost identical, it is evident that this difference is clearly attributed to the method used to represent soil stiffness using equivalent springs, as well as to the determination of the actual soil stiffness at the laboratory. It is interesting to notice that even though the hammer excitations were of low intensity and the induced soil shear strain subsequently small, the value of soil stiffness that was introduced in the numerical model was overestimated.

14.6   Conclusions

This paper presents an effort to comparatively assess the efficiency of numerical models to capture the effect of soil compliance on the predicted dynamic characteristics of bridge-foundation-soil systems. The study focuses on the case of the Metsovo bridge during the construction stage where measurements were made available and compared to the results of equivalent scaled systems tested in the laboratory. Due to the difficulties in constructing an actual concrete deck at a scale 1:100, an equivalent steel scaled structure was constructed in the laboratory presenting minimum deviation (as low as 6%) in terms of dynamic characteristics. The respective finite element model also successfully predicted the natural frequencies of the fixed scaled structure presenting a 2.12% average error. When the scaled structure was embedded in stabilized soil, a decrease was observed both experimentally and numerically for all considered modes. Three methods were adopted to simulate the soil compliance of the stabilized soil, namely: (a) a holistic method with 3D solid finite elements, (b) a 6+6 DOF springs method suggested by Kausel (1974) and Varun and Gazetas (2009) and (c) a 6-DOF spring method introduced by (Elsabee and Morray 1977). The average deviation between the identified and the numerically predicted natural frequencies range at all three methods between 11.3–14.5%, indicating that the stabilized soil's measured shear modulus was probably overestimated. Despite this, the influence of soil compliance was demonstrated by all numerical and experimental data thus highlighting the necessity of carefully considering soil compliance in the framework of structural health monitoring.

References

Basseville M, Benveniste A, Goursat M, Hermans L, Mevel L, Van der Auweraer H (2001) Output-only subspace-based structural identification: from theory to industrial testing practice. J Dyn Sys Meas Control 123(4):668–676

Bridgman PW (1931) Dimensional analysis, 2nd edn. Yale University Press, New Haven

Chaudhary MTA, Abe M, Fujino Y (2000) System identification of two base-isolated buildings using seismic records. J Struct Eng 126(10):1187–1195

Chaudhary M, Abe M, Fujino Y (2001) Identification of soil-structure interaction effects in base isolated bridges from earthquake records. Soil Dyn Earthq Eng 21:713–725

Crouse C, Hushmand B, Martin G (1987) Dynamic soil-structure interaction of single-span bridge. Earthq Eng Struct Dyn 15:711–729

DIN EN 459-1:2010–12, Building lime—Part 1: Definitions, specifications and conformity criteria

Elsabee F, Morray JP (1977) Dynamic behaviour of embedded foundations. Research Report R77–33 MIT

Kausel E (1974) Soil-forced vibrations of circular foundations on layered media. Research Report R74–11 MIT

Morassi A, Tonon S (2008) Dynamic testing for structural identification of a bridge. J Bridge Eng 13(6):573–585

Panetsos P, Ntotsios E, Liokos N, Papadimitriou C (2009) Identification of dynamic models of Metsovo (Greece) Bridge using ambient vibration measurements. In: ECCOMAS thematic conference on computational methods in structural dynamics and earthquake engineering (COMPDYN'09), Rhodes

Peeters B, De Roeck G (2001) Stochastic system identification for operational modal analysis: a review. J Dyn Sys Meas Control 123(4):659–667

Varun AD, Gazetas G (2009) A simplified model for lateral response of large diameter caisson foundations—Linear elastic formulation. Soil Dyn Earthq Eng 29(2):268–291

Werner SD, Beck JL, Levine MB (1987) Seismic response evaluations of Meloland road overpass using 1979 imperial valley earthquake records. Earthq Eng Struct Dyn 15:49–274

Wolf JP (1989) Soil-structure interaction analysis in time domain. Nucl Eng Des 11(3):381–393

Chapter 15
Quantification of the Seismic Collapse Capacity of Regular Frame Structures

David Kampenhuber and Christoph Adam

Abstract The aim of the present contribution is provide a global quantification of the effect of material deterioration on the seismic collapse capacity of regular frame structures vulnerable to the destabilizing effect of gravity loads (P-Delta effect). The record-to-record variability of the collapse capacity is studied using a set of ordinary earthquake records compiled in FEMA-P695. Median collapse capacities and their dispersions are presented in 3D in terms of contour plots as a function of the number of stories and a lateral stiffness quantification coefficient. Sets of structural parameters of generic frames are isolated, for which the collapse capacity of P-Delta sensitive structures is significantly reduced by material deterioration. From the outcomes of this study can be concluded that material deterioration affects the median and dispersion of the collapse capacity of P-Delta vulnerable multi-story frame structures in much the same manner as single-degree-of-freedom systems.

Keywords Collapse capacity · P-Delta effect · Earthquake excitation · Dispersion · Backbone curve deterioration

15.1 Introduction and Motivation

Sidesway collapse of a structural building subjected to severe earthquake excitation is the consequence of successive reduction of the lateral load carrying capacity resulting from stiffness and strength deterioration, and the global destabilizing effect of gravity loads acting through lateral displacements (P-Delta effect) (Krawinkler et al. 2009). In very flexible buildings the destabilizing effect of gravity loads may lead to a negative post-yield stiffness, and as a consequence, the structural collapse capacity might be exhausted at a rapid rate when the earthquake drives the structure into its inelastic range of deformation, even for stable hysteretic component behavior (Adam et al. 2004). In many other buildings the destabilizing effect of gravity loads

D. Kampenhuber · C. Adam (✉)
University of Innsbruck, Innsbruck, Austria
e-mail: christoph.adam@uibk.ac.at

© Springer International Publishing AG 2017
A.G. Sextos and G.D. Manolis (eds.), *Dynamic Response of Infrastructure to Environmentally Induced Loads*, Lecture Notes in Civil Engineering 2,
DOI 10.1007/978-3-319-56136-3_15

269

is negligible, and structural collapse is a result of component degradation only. The actual failure mechanism for this type of collapse is, thus, governed by the structural configuration and earthquake characteristics.

The collapse capacity of highly inelastic P-Delta sensitive single-degree-of freedom (SDOF) systems with non-deteriorating cyclic behavior has been previously studied rigorously in Adam and Jäger (2012a), Jäger (2012). For a set of representative ground motions, collapse capacity spectra have been derived, which quantify the collapse capacity as a function of the initial period of vibration, the normalized negative post-yield stiffness, and the viscous damping coefficient. Based on these spectra and on an equivalent single-degree-of-freedom system, an estimate of the median and dispersion of the collapse capacity of a P-Delta vulnerable non-deteriorating multi-degree-of-freedom (MDOF) frame structure can be determined (Jäger 2012; Adam and Jäger 2012b). The main assumptions of this so-called "collapse capacity spectrum methodology" are that the impact of material deterioration on seismic sidesway collapse of moment resisting frame structures is negligible compared to the destabilizing effect of gravity loads, and that sidesway collapse is governed by the fundamental mode.

Motivated by the work on solely P-Delta vulnerable systems (Adam and Jäger 2012a, b; Jäger 2012), Kampenhuber and Adam (2013) conducted a parametric study to assess the impact of material deterioration on the seismic collapse capacity of a set of SDOF systems with different hysteretic rules (without considering softening effects), varying the negative post-yield stiffness in a wide range. In Kampenhuber and Adam (2013) results of this study are presented in spectral form, plotting the median of the record-to-record dependent collapse capacity for a set of characteristic model parameters against the initial system period T. One illustrative outcome is shown in Fig. 15.1, where the left subfigure represents the relative collapse capacity of non-material deteriorating P-Delta sensitive "base case" systems (compare Adam and Jäger 2012a, b; Jäger 2012) for different negative post-yield stiffness ratios as a function of T. The negative post-yield stiffness ratio is characterized by the difference of the stability coefficient θ and the strain harding coefficient α, $\theta - \alpha$

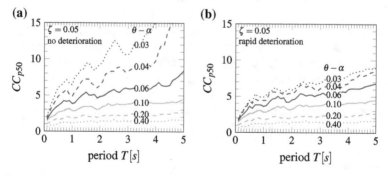

Fig. 15.1 Median collapse capacity spectra for P-Delta sensitive SDOF systems exhibiting **a** non-deteriorating and **b** rapid deteriorating bilinear cyclic behavior

(MacRae 1994). The right subfigure shows the corresponding outcomes for the combined material and P-Delta degrading counterpart system.

It is readily observed that for small $\theta - \alpha$ values up to 0.06 material deterioration reduces considerably the median collapse capacity of long period systems. In contrast, for pronounced negative post-yield stiffness ratios larger than 0.10 material deterioration has only a minor effect on the median collapse capacity. As shown in Kampenhuber and Adam (2013), for such P-Delta prone systems the underlying hysteretic cyclic material model does not affect significantly the collapse capacity.

Since the influence of material deterioration in combination with P-Delta only on the median collapse capacity of SDOF systems has been quantified (Kampenhuber 2013), the effect of material deterioration on the collapse capacity of P-Delta vulnerable MDOF structures needs to globally evaluated and clearly understood. The ultimate aim of this study is to reveal whether it is reasonable to extend the collapse capacity spectrum methodology for P-Delta vulnerable and material deteriorating multi-story frame structures. Therefore, in the present paper the relative global collapse capacity of a set of generic planar regular frame structures is discussed, evaluating the contributions of P-Delta and material deterioration on this quantity.

Note that here "deterioration" refers to the successive reduction of material/component quantities such as strength and unloading stiffness as a result of cyclic deformation. The degrading effect of P-Delta on the structure is not addressed with this term. Superscript ($\cdot^{wG}$) refers to quantities considering gravity loads (read: "with gravity"), whereas superscript ($\cdot^{woG}$) represents parameters disregarding gravity loads (read: "without gravity").

15.2   Structural Modeling Strategy

To draw meaningful conclusions, the need for generality of the results becomes a critical issue. Therefore, the considered structural frame models are not intended to represent a specific structure itself, but rather they should reveal the sensitivity of the collapse capacity to different modeling parameters. A so-called *generic* frame model satisfies these demands, because each characteristic structural parameter can be varied independently without affecting the others. Gupta and Krawinkler (1999) have shown that a single-bay generic frame is adequate to represent the global seismic response behavior also of a multi-bay frame. Consequently, in the present study a set of single-bay generic frame structures is utilized, based on models described in Medina and Krawinkler (2003). These structural models have already been used in other related studies such as in Jäger (2012), Adam and Jäger (2012b), Ibarra and Krawinkler (2005), Lignos and Krawinkler (2012).

15.2.1 General Model Properties

The utilized n story generic moment-resisting single-bay frame structures of uniform story height h, designed according to the weak beam—strong column design philosophy, are composed of rigid beams, elastic flexible columns, and inelastic rotational springs at the beam ends and the base according to a concentrated plasticity formulation. To each joint of the frames an identical point mass $m_i/2 = m_s/2, i = 1, \ldots, n$, and an identical gravity load is assigned. Figure 15.2a shows exemplarily the model of a generic four-story frame.

The targeted straight-line fundamental mode shape is the governing condition for adjusting the bending stiffness of the columns and the initial stiffness of the springs. Coefficient τ relates the fundamental period of the frame model without gravity loads, T_1^{woG}, and the number of stories of the frame structures, n, according to

$$T_1^{woG} = \tau n \tag{15.1}$$

and thus, quantifies the global lateral stiffness of the structure. As discussed in Medina and Krawinkler (2003), Ibarra and Krawinkler (2005), periods $T_1^{woG} = 0.10n$ and $T_1^{woG} = 0.20n$ are a reasonable lower and upper bound for moment-resisting frames, representing stiff and flexible structures, respectively. The strength of the rotational springs is tuned with the result that in a first mode pushover analysis without taking into account gravity loads yielding is initiated at all springs simultaneously. A predefined base shear coefficient $\gamma = V_{b,y}^{woG}/W$, i.e. the base shear without gravity loads at the onset of yielding $V_{b,y}^{woG}$ over the total structural weight $W = Mg$, $M = \sum_{i=1}^{n} m_s$, of 0.10 governs the magnitude of the yield strength (Medina and Krawinkler 2003).

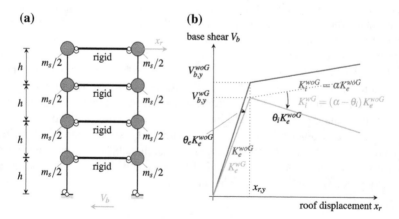

Fig. 15.2 **a** Mechanical model of a generic four-story frame. **b** Corresponding global pushover curves considering and disregarding gravity loads

Rayleigh type damping enforcing 5% viscous damping of the first mode and of that mode, where 95% of the total mass is exceeded, is considered. The corresponding damping matrix is proportional to the mass matrix and the current stiffness matrix.

15.2.2   Material Model

To each rotational spring a bilinear backbone curve is assigned. That is, a linear elastic branch of deformation is followed by a linear inelastic branch with reduced stiffness, characterized by the strain hardening coefficient α_s. In this study, α_s is the same for all springs in the considered MDOF models: $\alpha_s = 0.03$. The hysteretic response of the springs is assumed to be bilinear. Component deterioration is simulated with the modified Ibarra-Medina-Krawinkler deterioration model (Ibarra and Krawinkler 2005; Lignos and Krawinkler 2012; Ibarra et al. 2005). In this model, the prime parameter for cyclic stiffness and strength deterioration is the dissipated hysteretic energy, expressed in terms of the variable $\beta_{m,i}$:

$$\beta_{m,i}^{+/-} = \left(\frac{E_i}{E_{\text{total},m} - \sum_{j=1}^{i-1} E_j} \right)^{c_m} D^{+/-} \tag{15.2}$$

E_i represents the hysteretic energy dissipated in the ith inelastic excursion, $\sum_{j=1}^{i} E_j$ is the hysteretic energy dissipated in all previous excursions through loading in both positive and negative direction, and

$$E_{\text{total},m} = \Lambda_m M_y \tag{15.3}$$

denotes the hysteretic energy dissipation capacity, controlled by deterioration parameter Λ_m. Subscript m denotes the mode of deterioration. It is assumed that E_t is an inherent property of the components regardless of the loading history applied to the component (Lignos and Krawinkler 2011). Λ_m can be defined independently for acceleration reloading stiffness deterioration (then denoted as Λ_A), unloading stiffness deterioration (Λ_K), cyclic strength deterioration (Λ_S), and post-capping strength deterioration (Λ_C). Since the present study considers components with bilinear hysteretic behavior only, the acceleration reloading stiffness deterioration mode (Λ_A) does not exist. Furthermore, post-capping strength deterioration (Λ_C) is omitted, and has, thus, no influence on the results. Exponent c controls the rate of deterioration of the evaluated hysteretic parameter. According to Rahnama and Krawinkler (1993) c is a non-dimensional quantity between 1.0 and 2.0. Parameter D defines the decrease of rate of the cyclic deterioration in positive (D^+) and negative (D^-) loading direction, and can only be $0 \leq D^{+/-} \leq 1$. When the rate of cyclic deterioration is the same in both loading directions, then $D^{+/-} = 1$ (Lignos and Krawinkler 2012).

For instance, in the ith inelastic excursion strength deterioration is governed by

$$M_{y,i}^{+/-} = \left(1 - \beta_{S,i}\right) M_{y,i-1}^{+/-} \tag{15.4}$$

$M_{y,i}$ denotes the yield moment after the ith excursion, and $M_{y,i-1}$ is the yield moment before this excursion. $\beta_{S,i}$ is the corresponding strength deterioration coefficient as defined in Eq. 15.2.

These deterioration rules are assigned to the rotational springs of the generic frame models defined upfront, where each rotational spring of a specific structure exhibits the same deterioration parameters. For the sake of simplicity, unloading stiffness deterioration and cyclic strength deterioration parameters Λ_K and Λ_S, respectively, are assumed to be equal, $\Lambda_S = \Lambda_K$. In Lignos and Krawinkler (2011, 2012) a database of experimental cyclic studies on steel components is complied, which allowed to identify the significant parameters affecting the cyclic moment-rotation relationship at plastic hinge regions in beams. Based on this study, in the present contribution three different levels of deterioration, representing slow, medium, and rapid deterioration, are utilized.

15.2.3 P-Delta Effect

The P-Delta effect on multi-story frame structures is quantified through two base shear-roof drift (V_b–x_r) relations (referred to as global pushover curves), resulting from two different first mode pushover analyses as shown in Fig. 15.2b (Adam and Jäger 2012b). In the first pushover analysis in the initial step the gravity loads are applied to the model, whereas the second pushover analysis is conducted without gravity loads. For the present generic frame structures both global pushover curves are bilinear, because at all springs yielding is initiated simultaneously. As depicted in Fig. 15.2b, the P-Delta effect leads for a given roof displacement to a reduction of the base shear, and thus, to an apparent reduction of the lateral stiffness. It has already been shown by Medina and Krawinkler (2003) that for multi-story frame structures this stiffness reduction is different in the elastic and inelastic deformation branch of the global pushover curve. Consequently, additionally to the lateral global hardening coefficient α, an elastic and an inelastic stability coefficient, θ_e and θ_i, respectively, which characterizes the magnitude of the stiffness reduction in these deformation branches, can be identified from the pushover curves,

$$\theta_e = 1 - \frac{K_e^{wG}}{K_e^{woG}} \quad , \quad \theta_i = \alpha - \frac{K_i^{wG}}{K_e^{woG}} \tag{15.5}$$

Table 15.1   Summary of basic model parameters of the considered frame structures

Variable	Description	Range of values
n	Number of stories	1–20
τ	Global stiffness quantification factor	$0.10, 0.12, 0.14, 0.16, 0.18, 0.20$
$\Lambda_K = \Lambda_S$	Deterioration parameter representing no, slow, medium and rapid material deterioration, respectively	$\infty, 2.0, 1.0, 0.5$
α_s	Strain hardening coefficient of rotational springs	0.03
$\theta_i - \alpha$	Negative post-yield stiffness ratio	$0.03, 0.04, 0.05, 0.06, 0.10, 0.20, 0.30, 0.40$
γ	Base shear coefficient at yield	0.10
ζ	Damping coefficient for Rayleigh damping	0.05

Here, K_e^{wG} and K_e^{woG} denotes the global lateral elastic stiffness considering and disregarding P-Delta, respectively, and K_i^{wG} and K_i^{woG} represents the global lateral inelastic "stiffness" considering and disregarding P-Delta, respectively, compare also with Fig. 15.2b. In general the lateral global hardening coefficient α is not the same as the strain hardening coefficient α_s assigned to each rotational spring. In the present study the difference $\theta_i - \alpha$, i.e. the normalized slope of the inelastic deformation branch (also referred to as negative post-yield stiffness ratio), is the target variable that characterizes the P-Delta effect on the frame structure (compare with Fig. 15.2b) (Adam et al. 2017). Thus, for each structure with different properties but the same $\theta_i - \alpha$ value, the ratio of total gravity load over dead weight is in general different. This is a contrast to previous studies (Jäger 2012; Adam and Jäger 2012b; Ibarra and Krawinkler 2005; Adam et al. 2017), where the P-Delta effect has been characterized by a constant gravity load coefficient ν.

15.2.4   Generic Model Parameters

All predefined basic model parameters of the considered generic frame structures are summarized in Table 15.1. Considering each possible combination of parameters leads in total to 3840 different frame structures to be investigated.

15.3   Global Seismic Collapse Capacity

In a parametric study, the global seismic collapse capacity of 3840 multi-story frame structures is determined performing incremental dynamic analysis (IDA) (Vamvatsikos and Cornell 2002). In an IDA nonlinear response history analyses are conducted repeatedly, increasing in each subsequent run the ground motion intensity. As outcome, an appropriate measure of the intensity (IM) of the earthquake record is plotted against the engineering-demand-parameter (EDP). The analysis is stopped when the EDP satisfies a certain failure criterion that may correspond to structural collapse (Adam and Ibarra 2015). The corresponding intensity of the ground motion is referred to as global collapse capacity of the building subjected to this specific ground motion record. In the present study, global collapse is assumed to be indicated when an incremental increase of the intensity leads to an unbounded structural response. A general overview of seismic collapse assessment is provided in Adam and Ibarra (2015).

In the present contribution the 5% damped spectral pseudo-acceleration S_a at the fundamental structural period T_1^{wG}, normalized with respect to the base shear coefficient γ^{wG} at yield and acceleration of gravity g, serves as characteristic relative intensity measure IM of an earthquake record (Tsantaki 2014),

$$IM = \frac{S_a(T_1^{wG}, \zeta = 5\%)}{g\gamma^{wG}} \quad , \quad \gamma^{wG} = \frac{V_{b,y}^{wG}}{W} \tag{15.6}$$

As discussed in Tsantaki (2014), it is beneficial to utilize in this definition of the relative seismic intensity the structural quantities affected by gravity. The relative collapse capacity is therefore given by

$$CC = IM|_{collapse} \tag{15.7}$$

To capture the inherent record-to-record variability of the collapse capacity, in this study the 44 ordinary ground motions compiled in the far-field set of the FEMA P-695 report FEMA (2009) are used, and subsequently median and dispersion of the 44 corresponding collapse capacities of each structure are evaluated.

As discussed in Shome et al. (1998), it is reasonable to approximate the seismic record-dependent collapse capacity variability by a log-normal distribution (Ibarra and Krawinkler 2005). For P-Delta sensitive structures this approximation has been confirmed in Adam and Jäger (2012a, b). A log-normal distribution is characterized by a measure of central tendency ($\bar{\mu}$) and a measure for dispersion (σ^2) (Limpert et al. 2001). These two quantities can be related to the median, 16th and 84th percentiles of the individual record-depending collapse capacities CC_i, $i = 1, ..., 44$, (referred to as CC^{P50}, CC^{P16}, and CC^{P84}, respectively) (Jäger 2012):

$$s^* = \sqrt{\frac{CC^{P84}}{CC^{P16}}} \qquad \sigma = \ln s^* \qquad \bar{\mu} = \ln CC^{P50} \tag{15.8}$$

The "counted" statistical value s^* should be utilized when for several records of a set no collapse is attained. The use of s^* allows the derivation of analytical fragility curves based on the 16th, 50th, and 84th percentile of the collapse capacities. Dispersion measure s^* may exclude possible outliers that commonly correspond to larger values of collapse capacities than those obtained from the 84th percentile (Tsantaki 2014). Alternatively, parameter β,

$$\beta = \sqrt{\sum_{i=1}^{N=44} \frac{(\ln CC_i - \mu_{\ln CC})^2}{N-1}} \qquad (15.9)$$

can be utilized as characteristic parameter of the dispersion. With β the deviation through the whole range of the N collapse capacities including possible outliers is directly calculated, and thus, referred to as "computed" statistical quantity (Tsantaki 2014). In Eq. 15.9, $\mu_{\ln CC_i}$ is the mean of the natural logarithm of the collapse capacity.

In this article, the dispersion of the record-to-record variability of the collapse capacity is quantified through parameter β.

15.4   Numerical Results

To reveal systematically the influence of material deterioration on the collapse capacity with respect to the initial lateral stiffness (and thus, with respect to the first mode period, see Eq. 15.1) and the vulnerability to P-Delta expressed through $\theta_i - \alpha$, the presentation of the results is organized as follows.

- Median collapse capacities and their dispersions are presented three-dimensionally in terms of contour plots as a function of the number of stories and a lateral stiffness quantification coefficient.
- The results of each figure are based on the same material model. That is, in Figs. 15.3 and 15.4 the outcomes of frames with assigned *non*-deteriorating material model are shown. These results are referred to as "base case". The results of Figs. 15.5 and 15.8 include the effect of *slow* material deterioration, Figs. 15.6 and 15.9 depict the results based on *medium* material deterioration, and Figs. 15.7 and 15.10 correspond to *rapid* deterioration of strength and stiffness.
- Superscripts *"bc"* and *"det"* denote results of non-deteriorating base case frame models and material deteriorating frames, respectively.
- Each subfigure of a figure refers to a certain stiffness quantification factor τ, starting with $\tau = 0.10$ (subfigure (a)), and step-wise increasing τ by 0.02 for each subsequent subfigure up to $\tau = 0.20$ for laterally flexible structures (subfigure (f)).
- All results considering a deteriorating material model are normalized with respect to the corresponding outcomes of the base case (i.e., for the median collapse capacity $CC_{p50}^{det}/CC_{p50}^{bc}$, and for the dispersion β^{det}/β^{bc}).
- The value range represented through colorbars, which are depicted on the right side of each figure, is the same for all figures, but different for absolute and normalized values. Therefore, the outcomes of different parameter configurations can be directly compared.

278 D. Kampenhuber and C. Adam

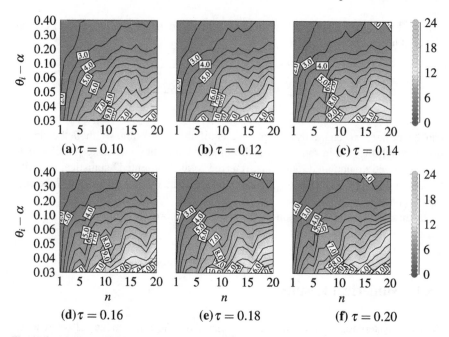

Fig. 15.3 Median collapse capacity (CC_{p50}) of non-deteriorating P-Delta vulnerable frame structures (base case) for different stiffness quantification coefficients τ

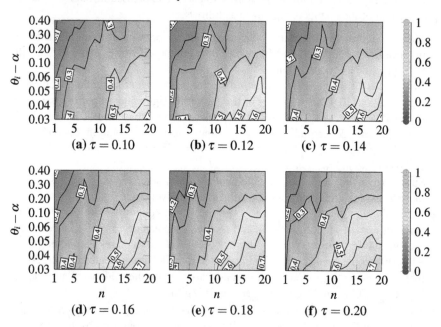

Fig. 15.4 Measure of dispersion (β) of non-deteriorating P-Delta vulnerable frame structures (base case) for different stiffness quantification coefficients τ

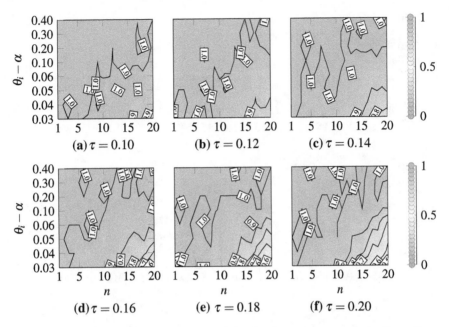

Fig. 15.5 Median collapse capacity of deteriorating frame structures over the median collapse capacity of the non-deteriorating counterparts ($CC_{p50}^{det}/CC_{p50}^{bc}$) for different quantification coefficients τ and *slow* component deterioration

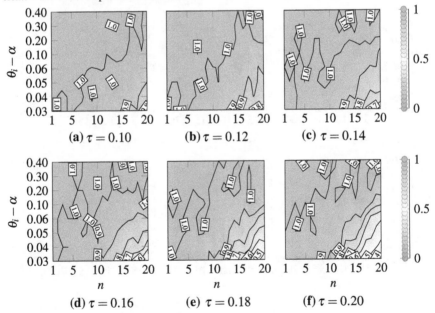

Fig. 15.6 Median collapse capacity of deteriorating frame structures over the median collapse capacity of the non-deteriorating counterparts ($CC_{p50}^{det}/CC_{p50}^{bc}$) for different quantification coefficients τ and *medium* component deterioration

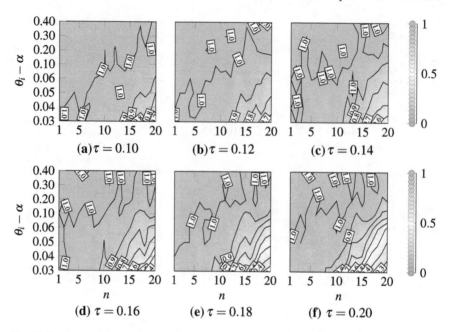

Fig. 15.7 Median collapse capacity of deteriorating frame structures over the median collapse capacity of the non-deteriorating counterparts ($CC_{p50}^{det}/CC_{p50}^{bc}$) for different quantification coefficients τ and *rapid* component deterioration

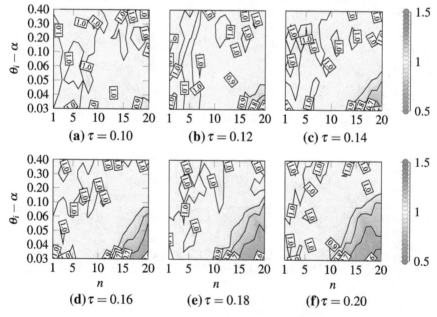

Fig. 15.8 Measure of dispersion of deteriorating frame structures over the measure of dispersion of the non-deteriorating counterparts (β^{det}/β^{bc}) for different quantification coefficients τ and *slow* component deterioration

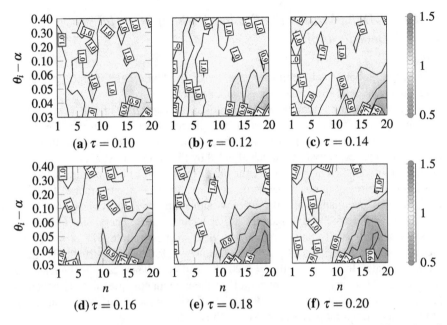

Fig. 15.9 Measure of dispersion of deteriorating frame structures over the measure of dispersion of the non-deteriorating counterparts (β^{det}/β^{bc}) for different quantification coefficients τ and *medium* component deterioration

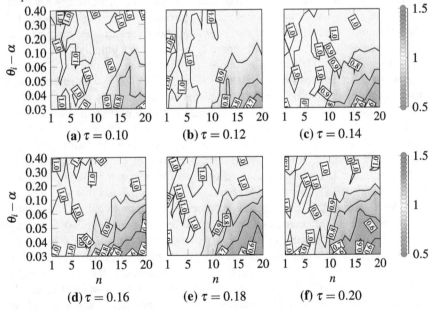

Fig. 15.10 Measure of dispersion of deteriorating frame structures over the measure of dispersion of the non-deteriorating counterparts (β^{det}/β^{bc}) for different quantification coefficients τ and *rapid* component deterioration

15.4.1 P-Delta Vulnerable Frames Without Material Deterioration

At first the global seismic collapse capacity of the "base case" frame structures, where material deterioration is disregarded, is discussed.

In Fig. 15.3 the relative median collapse capacity of all considered non-deteriorating frames (in total 960) is depicted. It can be seen that for structures with a large negative post-yield stiffness ratio, i.e. $\theta_i - \alpha = 0.20$ to 0.40, the median collapse capacity is not much affected by the number of stories and stiffness quantification coefficient τ (compare the dark blue areas of the subfigures). In contrast, in structures of minor P-Delta vulnerability, i.e., $\theta_i - \alpha$ in the range of 0.03 to 0.05, the median collapse capacity varies with the number of stories n and the structural lateral stiffness characterized by τ. For instance, for a lateral stiff ($\tau = 0.10$) 20-story building with small negative post-yield stiffness ratio of $\theta_i - \alpha = 0.03$ the median collapse capacity is $CC_{p50} = 14.94$. However, for the same structural configuration but laterally flexible with $\tau = 0.20$, the median collapse capacity increases to $CC_{p50} = 26.27$. In general, the magnitude of the median collapse capacity becomes larger with increasing number of stories and flexibility, and thus, with increasing fundamental structural period, compare with Eq. 15.1.

For the same structural configurations in Fig. 15.4 the corresponding measure of the record-to-record collapse capacity dispersion β according to Eq. 15.9 is depicted. The outcomes show that β is almost proportional to the building flexibility τ and the number of stories n, and it increases as τ and n increases. β is inversely proportional to the vulnerability to P-Delta expressed by $\theta_i - \alpha$, and decreases as $\theta_i - \alpha$ increases.

For instance, for a stiff one-story frame ($\tau = 0.10$, $n = 1$) highly vulnerable to P-Delta ($\theta_i - \alpha = 0.40$) parameter β is 0.09 (see Fig. 15.4a). In contrast, a 20-story frame with the same negative post-yield stiffness ratio $\theta_i - \alpha = 0.40$ and the same stiffness quantification factor $\tau = 0.10$ the dispersion increases by a factor of 3.5 to $\beta = 0.32$. A stiff one-story structure with small negative post-yield stiffness ratio ($\tau = 0.10$, $\theta_i - \alpha = 0.03$, $n = 1$) exhibits a collapse capacity dispersion of $\beta = 0.20$. However, increasing the number of stories to 20, which means that also the fundamental period of vibration is 20 times larger, increases the dispersion parameter to $\beta = 0.63$, which is twice as large as the dispersion for the corresponding previously considered highly P-Delta vulnerable model. This outcome shows that with increasing P-Delta vulnerability the structure becomes more prone to collapse, and thus, the collapse variability decreases. The record-to-record induced collapse capacity dispersion β of a very flexible ($\tau = 0.20$) highly P-Delta vulnerable ($\theta_i - \alpha = 0.40$) frame is 0.37, compared to 0.10 of the one-story counterpart model (see Fig. 15.4f). It is should be, however, kept in mind that the period of vibration of the one-story frame model is 20 times smaller than the fundamental period of the 20-story structure. In a flexible structure with $\tau = 0.20$ and low P-Delta vulnerability ($\theta_i - \alpha = 0.03$) of one-story dispersion β is 0.29, and of 20-stories β is 0.83.

15.4.2   P-Delta Vulnerable Frames Exhibiting Material Deterioration

To assess the effect of material deterioration on the median collapse capacity (CC_{p50}) and on the record-to-record dependent collapse capacity dispersion parameter β, three different speeds of component deterioration as listed in Table. 15.1 are taken into account. The deterioration speed effects the choice of the deterioration parameter Λ (Ibarra and Krawinkler 2005; Lignos and Krawinkler 2012).

In Figs. 15.5, 15.6 and 15.7 the derived median collapse capacities CC_{p50}^{det} are normalized with respect to the outcomes CC_{p50}^{bc} of the base case. In particular, the collapse capacity ratios $CC_{p50}^{det}/CC_{p50}^{bc}$ based on *slow* (Fig. 15.5), *medium* (Fig. 15.6), and *rapid* (Fig. 15.7) material deterioration are presented.

Figures 15.5a, 15.6a, and 15.7a visualize the effect of material deterioration for three deterioration speeds on the collapse capacity ratios of *stiff* structures ($\tau = 0.10$). It is readily observed that for low- to mid-rise frame structures ($n \leq 10$) the median collapse capacity ratio is close to 1.0, which means that the median collapse capacity of the corresponding P-Delta vulnerable structures is in general not significantly affected by material deterioration. Only if for high-rise frames the negative post-yield stiffness ratio $\theta_i - \alpha$ is small, deterioration becomes more important in reducing the collapse capacity, in particular for rapid material deterioration (Fig. 15.7a). For instance, for a 20-story frame with $\theta_i - \alpha = 0.03$ and rapid material deterioration the ratio $CC_{p50}^{det}/CC_{p50}^{bc}$ is 0.68, which means that material deterioration reduces the median collapse capacity by 32%. From the results it can be concluded that collapse of the considered stiff structures up to 10 stories is primarily governed by P-Delta. Figures 15.5f, 15.6f, and 15.7f representing flexible frame structures ($\tau = 0.20$) show a different picture. It can be seen that material deterioration plays a more prominent role compared to stiff structures. For those *flexible* frames a significant influence of material deterioration on the median collapse capacity is observed if P-Delta is less pronounced (i.e. $\theta_i - \alpha < 0.06$) and the number of stories is larger than 10. Here, the median collapse capacity ratio $CC_{p50}^{det}/CC_{p50}^{bc}$ of the 20-story frame with $\theta_i - \alpha = 0.03$ is 0.23, which is about 66% smaller than for the stiff counterpart structure. The entire set of frame structures exhibiting stiff and flexible lateral stiffness has in common that the collapse capacity is only negligibly influenced by material deterioration if the structures are highly vulnerable to P-Delta, i.e. the negative post-yield stiffness ratio $\theta_i - \alpha$ is larger than 0.20. Consequently, in this parameter domain the collapse capacity ratios shown in Figs. 15.5, 15.6, and 15.7 are close to 1.0.

Figures 15.8, 15.9 and 15.10 represent the measure of collapse capacity dispersion β through the ratios β^{det}/β^{bc} for *slow* (Fig. 15.8), *medium* (Fig. 15.9), and *rapid* (Fig. 15.10) material deterioration. The outcomes depicted in these figures reflect the trend of β with respect to the number of stories and the negative post-yield stiffness ratios as observed for non-deteriorating structures (Fig. 15.4). This is quite welcome, because it indicates a decrease of the dispersion in those parameter domains, where the dispersion of the non-deteriorating counterpart structures has its maximum. However, it should be kept in mind that in these domains the collapse capacity decreases due to the fact that material deterioration makes the structures more vulnerable to collapse (compare Figs. 15.5, 15.6 and 15.7). For instance, a flexible ($\tau = 0.20$) 20-story building with small $\theta_i - \alpha$ of 0.03 exhibits an absolute measure of dispersion of $\beta^{bc} = 0.83$. β^{bc} value becomes smaller for slow deterioration, i.e., $\beta^{det(slow)} = 0.43$, represented by a dispersion ratio of $\beta^{det}/\beta^{bc} = 0.52$, see Fig. 15.8f. For medium and rapid deteriorating systems the absolute dispersion decreases in both cases to $\beta^{det(medium,rapid)} = 0.44$, which corresponds to a dispersion ratio of $\beta^{det}/\beta^{bc} = 0.53$, see Figs. 15.9f and 15.10f. Comparing the subfigures of Figs. 15.8, 15.9 and 15.10 reveals that β is not considerably affected by the deterioration speed itself, but in general affected by material deterioration, in particular for less P-Delta vulnerable, high-rise buildings. In low-rise buildings the material deterioration process does not change the dispersion. In these cases the dispersion of non-deteriorating counterpart frames is generally low. It is furthermore observed that with increasing structural flexibility the reduction of the dispersion becomes larger, in particular for high-rise frames with small negative post-yield stiffness ratio.

15.5 Summary and Conclusions

In this contribution the influence of material deterioration on the global seismic collapse capacity of multi-story P-Delta vulnerable frame structures has been quantified. The results from the parametric study of in total 3840 frames can be summarized as follows.

- The median collapse capacity of P-Delta sensitive low- to mid-rise frame structures is not significantly affected by material deterioration, in particular if the structure is laterally stiff.
- In flexible frames material deterioration leads to a significant decrease of the median collapse capacity if the negative post-yield stiffness due to P-Delta is less pronounced (i.e., the negative post-yield stiffness ratio $\theta_i - \alpha \leq 0.06$) and the number of stories n is larger than 10.
- Stiff and flexible frame structures have in common that the median collapse capacity is only negligibly influenced by material deterioration if the structures are highly vulnerable to P-Delta (i.e., $\theta_i - \alpha \geq 0.20$).
- Material deterioration results in a decrease of the collapse capacity dispersion in those parameter domains, where the dispersion of non-deteriorating frames has its maximum (i.e., $n \geq 10, \theta_i - \alpha \leq 0.06$).
- The collapse capacity dispersion of high-rise buildings with small negative lateral post-yield stiffness ratio (i.e., $n \geq 10, \theta_i - \alpha \leq 0.06$) is significantly affected by material deterioration, however, not by the underlying deterioration speed.
- In low-rise buildings the collapse capacity dispersion is not considerably affected by material deterioration.

From the derived results for P-Delta vulnerable multi-story frames it can be concluded that material deterioration reduces the relative seismic collapse capacity of systems with long periods and small P-Delta induced negative post-yield stiffness ratio. This behavior has been already observed in single-degree-of-freedom systems. Since in the single-degree-of-freedom and the multi-degree-of-freedom domain the collapse capacity exhibit the same trend, an incorporation of material deterioration into the collapse capacity spectrum methodology seems reasonable.

Acknowledgements This work was supported by the Austrian Ministry of Science BMWF as part of the UniInfrastrukturprogramm of the Focal Point Scientific Computing at the University of Innsbruck.

References

Adam C, Ibarra LF, Krawinkler H (2004) Evaluation of P-delta effects in non-deteriorating MDOF structures from equivalent SDOF systems. In: Proceedings of the 13th world conference on earthquake engineering (13WCEE), Vancouver, B.C., Canada, 1–6 Aug, Paper No. 3407

Adam C, Ibarra LF (2015) Seismic collapse assessment. In: Beer M, Kougioumtzoglou IA, Patelli E, Siu-Kui Au I (eds) Earthquake engineering encyclopedia, vol 3, Springer, Berlin Heidelberg, pp 2729–2752. doi:10.1007/978-3-642-36197-5_248-1, Online ISBN 978-3-642-36197-5

Adam C, Jäger C (2012a) Seismic collapse capacity of basic inelastic structures vulnerable to the P-delta effect. Earthq Eng Struct Dyn 41:775–793

Adam C, Jäger C (2012b) Simplified collapse capacity assessment of earthquake excited regular frame structures vulnerable to P-delta. Eng Struct 44:159–173

Adam C, Kampenhuber D, Ibarra LF (2017) Optimal intensity measure based on spectral acceleration for P-delta vulnerable deteriorating frame structures in the collapse limit state. Bull Earthq Eng. doi:10.1007/s10518-017-0129-3

FEMA-P695 (2009) Quantification of building seismic performance factors. Federal Emergency Management Agency (FEMA), Washington, DC

Gupta A, Krawinkler H (1999) Prediction of seismic demands for SMRFs with ductile connections and elements. Report No. SAC/BD-99/06. SAC background document

Ibarra LF, Krawinkler H (2005) Global collapse of frame structures under seismic excitations. Report No. 152, The John A. Blume Earthquake Engineering Research Center, Department of Civil and Environmental Engineering, Stanford University, Stanford, CA

Ibarra LF, Medina RA, Krawinkler H (2005) Hysteretic models that incorporate strength and stiffness deterioration. Earthq Eng Struct Dyn 34:1489–1511

Jäger C (2012) The collapse capacity spectrum method. A methodology for rapid assessment of the collapse capacity of inelastic frame structures vulnerable to P-delta subjected to earthquake excitation (in German). PhD thesis, University of Innsbruck, 2012

Kampenhuber D, Adam C (2013) Vulnerability of collapse capacity spectra to material deterioration. In: Adam C, Heuer R, Lenhardt W, Schranz C (eds) Proceedings of the vienna congress on recent advances in earthquake engineering and structural dynamics 2013 (VEESD 2013), 28–30 Aug, Vienna, Austria, Paper No. 424, 2013

Krawinkler H, Zareian F, Lignos D, Ibarra LF (2009) Prediction of collapse of structures under earthquake excitations. In: Papadrakakis, M, Lagaros ND, Fragiadakis M (eds) 2nd international conference on computational methods in structural dynamics and earthquake engineering (COMPDYN 2009), 22–24 June, Rhodes, Greece, CD–ROM paper, Paper No. CD449, 19 pp

Lignos D, Krawinkler H (2011) Deterioration modeling of steel components in support of collapse prediction of steel moment frames under earthquake loading. J Struct Eng 137(11):1291–1302

Lignos D, Krawinkler H (2012) Sidesway collapse of deteriorating structural systems under seismic excitations. Report No. 177, The John A. Blume Earthquake Engineering Research Center, Department of Civil and Environmental Engineering, Stanford University, Stanford, CA

Limpert E, Stahel WA, Abbt M (2001) Log-normal distributions across the sciences: keys and clues. BioScience 51:341–352

MacRae GA (1994) P-Δ effects on single-degree-of-freedom structures in earthquakes. Earthq Spectra 10:539

Medina RA, Krawinkler H (2003) Seismic demands for nondeteriorating frame structures and their dependence on ground motions. Report No. 144, The John A. Blume Earthquake Engineering Research Center, Department of Civil and Environmental Engineering, Stanford University, Stanford, CA

Rahnama M, Krawinkler H (1993) Effects of soft soil and hysteresis model on seismic demands. Report No. 108, The John A. Blume Earthquake Engineering Research Center, Department of Civil and Environmental Engineering, Stanford University, Stanford, CA

Shome N, Cornell CA, Bazzurro P, Carballo JE (1998) Earthquakes, records, and nonlinear responses. Earthq Spectra 14(3):469–500

Tsantaki S (2014) A contribution to the assessment of the seismic collapse capacity of basic structures vulnerable to the destabilizing effect of gravity loads. PhD thesis, University of Innsbruck

Vamvatsikos D, Cornell CA (2002) Incremental dynamic analysis. Earthq Eng Struct Dyn 31:491–514

CPSIA information can be obtained
at www.ICGtesting.com
Printed in the USA
LVHW01*1509310718
585363LV00001B/86/P